AN INTRODUCTION TO THE BASICS
OF RELIABILITY AND RISK ANALYSIS

SERIES ON QUALITY, RELIABILITY AND ENGINEERING STATISTICS

Series Editors: M. Xie *(National University of Singapore)*
T. Bendell *(Nottingham Polytechnic)*
A. P. Basu *(University of Missouri)*

Published

Series in Quality, Reliability and Engineering Statistics Vol. 13

AN INTRODUCTION TO THE BASICS OF RELIABILITY AND RISK ANALYSIS

Enrico Zio

Polytechnic of Milan, Italy

 **World Scientific**

NEW JERSEY · LONDON · SINGAPORE · BEIJING · SHANGHAI · HONG KONG · TAIPEI · CHENNAI

Published by

World Scientific Publishing Co. Pte. Ltd.

5 Toh Tuck Link, Singapore 596224

USA office: 27 Warren Street, Suite 401-402, Hackensack, NJ 07601

UK office: 57 Shelton Street, Covent Garden, London WC2H 9HE

Library of Congress Cataloging-in-Publication Data
Zio, Enrico.
    An introduction to the basics of reliability and risk analysis / Enrico Zio.
    p. cm. -- (Series on quality, reliability & engineering statistics ; v. 13)
    Includes bibliographical references.
    ISBN-13 978-981-270-639-3
    ISBN-10 981-270-639-9
    1. Reliability (Engineering) 2. Risk assessment.
    TA169 .Z56 2007
    620'.00452--dc22

                                                2007278630

British Library Cataloguing-in-Publication Data
A catalogue record for this book is available from the British Library.

First published 2007
Reprinted 2010, 2012

Printed in Singapore by World Scientific Printers.

To my life team members: Giorgia, Aurora, Cecilia and Matteo
To my work team members: my students and collaborators

Milano, 21 December 2006

About the Book

This book introduces the principal concepts and issues related to the safety of modern industrial activities and presents the classical techniques for reliability analysis and risk assessment used in the current practice. It is aimed at providing an organic view of the subject.

The contents of the book comprise: *i)* a basic illustration of some methods of system analysis commonly used in practice for the identification of the hazards associated to industrial plants and processes; *ii)* a review of the basics of probability theory, tailored to its application to reliability analysis and risk assessment; *iii)* an overview of the basics of reliability, availability and maintainability applied to standard system configurations, such as series, parallel, stand-by and others; *iv)* a presentation of the fault tree and event tree analysis methods, which constitute powerful tools widely used in practice for the reliability and risk assessment of complex systems; *v)* a review of the statistical methods for the estimation of failure rates; *vi)* a sketch of some modelling techniques of reliability growth and prediction.

The book can serve as any senior undergraduate or post-graduate university course on the subject or as reference for the initiation of young researchers to the field. In this view, several numerical examples are provided when appropriate, as guide for the comprehension.

About the Author

Enrico Zio (BS in nuclear engineering., Politecnico di Milano, 1991; MSc in mechanical engineering., UCLA, 1995; PhD, in nuclear engineering., Politecnico di Milano, 1995; PhD, in nuclear engineering., MIT, 1998) is a full professor of Nuclear Engineering and Dean of the Graduate School of the Politecnico di Milano, Italy. He holds a course on *Computational methods for safety and risk analysis* at the Politecnico di Milano and has served as lecturer at various Master and PhD programs in Italy and abroad.

He has served as Vice-Chairman of the European Safety and Reliability Association, ESRA and as Editor-in-Chief of the International journal Risk, Decision and Policy. He is member of the editorial boards of two recognized international scientific journals and has been involved in the organization of various international conferences, in the field of Safety and Reliability.

His research interests are: analysis of the reliability, safety and security of complex systems under stationary and dynamic operation, particularly by Monte Carlo simulation methods and cellular automata; development of soft computing techniques (neural networks, fuzzy logic, genetic algorithms) for safety and reliability applications, system monitoring, fault diagnosis and optimal design. He is co-author of one international book on Monte Carlo simulation applied to reliability and risk analysis and of more than 100 papers on international journals.

Contents

1

Introduction

The necessity of expertise for tackling the complicated and multidisciplinary issues of reliability and risk analysis has slowly permeated into all engineering applications, with risk analysis and management gaining a relevant role both as a tool in support of plant design and operation, and as an indispensable means for emergency planning in accidental situations.

Failure is an unavoidable phenomenon in all technological products and systems. From the scientific and engineering point of view, the investigation of the uncertain and 'obscure' domain of failures entails the exploration of the functional and physical limits of systems, in an effort to understand how, why and when a device may not function properly. In this respect, the required approach is complementary to the traditional engineering viewpoint which focuses on how and when a machine functions in an optimal way.

Whatever particular failure one is considering, proper control and management of it become essential. Areas of application which involve failure-oriented and failure-driven aspects are Reliability, Availability, Maintainability, Safety (RAMS), Risk, Quality control (QC), Fault Detection and Identification (FDI), security and others. As such, failure analysis presents a strong connotation of multi-disciplinarity which significantly adds to its inherent difficulty. Hence, these failure-oriented disciplines have become more and more important and closely connected so as to require an integrated view. This entails the acquisition of appropriate modeling and analysis tools as complement to the basic and specific engineering knowledge for the technological area of application.

The present lecture notes draw from the specialized literature to address the above issues related to the safety of modern industrial activities and illustrate the classical techniques available for the evaluation, the management and the control of the associated risks. The motivation behind the effort of editing such notes derives from the need to offer a more organic view of the subject to the students who are attending my courses. In this sense, the contents are limited to the topics I teach in the classroom and are thus certainly not exhaustive of the very extensive subject of reliability and risk analysis and surely lacking in many ways. In any case, I believe that they can be of use in any senior undergraduate university course on the subject or as basis for the initiation of young researchers to the field. To this aim, several numerical examples are provided when appropriate, for ease of understanding.

Enrico Zio
Milano, December 2006

2

Basic Concepts of Safety and Risk Analysis

2.1 A qualitative definition of risk

The subject of risk has become very popular in the last few years and is much talked about at all levels of industry. We shall first give a definition of risk in qualitative terms and then translate it in quantitative terms [5] in the following Section.

A first, intuitive observation comes from the fact that there is *risk* if there exists a potential source of *damage*, or *hazard*. When a hazard exists, e.g. posed by a system which in certain conditions may cause undesired consequences, *safeguards* are typically devised to prevent the occurrence of such hazardous conditions and its associated undesired consequences. However, the presence of a hazard does not suffice itself to define a condition of risk. Indeed, inherent in the latter there is the *uncertainty* that the hazard translates from potential to actual damage. Thus, the notion of risk involves some kind of loss or damage that might be received and the uncertainty of its transformation in an actual loss or damage:

$$\text{Risk} = \text{Damage} + \text{Uncertainty}$$

This qualitative analysis is reflected in the various Dictionary-definitions of risk, such as 'possibility of loss or injury and the degree of probability of such loss'.

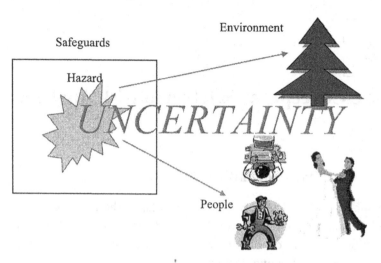

Fig. 2.1: The main components of the concept of risk

2.2 A quantitative definition of risk

Let x and p denote a given damage and the probability of receiving such damage, respectively. From a quantitative point of view, it is common to define a measure of the associated risk R as:

$$R = x \cdot p \qquad\qquad (2.1)$$

In practice, often, the perception of risk is such that the relevance given to the damaging consequences x is far greater than that given to its probability of occurrence p so that Eq. (2.1) is slightly modified to:

$$R = x^k \cdot p \qquad \text{with } k > 1 \qquad\qquad (2.2)$$

By so doing, numerically larger values of risk are associated to larger consequences.

When considering complex systems, the above quantitative definitions must be extended to account for the fact that typically more than one undesirable event exists. With n undesirable events associated with the operation of a given system, Eq. (2.1) is usually extended to the following definition of *composite risk* which accounts for all hazards present, in an integral way:

$$R = \sum_{i=1}^{n} x_i \cdot p_i \qquad (2.3)$$

and similarly for Eq. (2.2).

The quantitative definitions of risk in Eqs. (2.1), (2.2), (2.3) are however little informative for the purposes of risk analysis, management and regulation. Suppose you were considering two different systems A and B of equal risk $R_A = R_B$ as defined by (2.1). Let the risk of A be due to a potentially large consequence x_A occurring with small probability p_A and vice versa for the risk of B. Then, if we wish to intervene on the design, operation and regulation of the two systems in order to reduce the associated risks, we act differently knowing the different natures of the risk in the two cases. To reduce R_A we would implement emergency systems which mitigate the accident (*mitigation*) and containment systems which limit its consequences to the outside environment (*protection*); on the contrary, if we were to reduce R_B we would allocate additional redundancies and improve the reliability of the system components so as to reduce the probability of an accident (*prevention*). Thus, if we simply know the value of R, we may not be effective in reducing it by limiting its probability part or by mitigating its consequences; hence, the importance of keeping separate the constituents of risk: scenarios, p and x. Note also how the key concepts of the *defense-in-depth* approach, i.e. *prevention, mitigation, protection*, come into play in the management of risk.

The situation is even worse in the case of the composite risk of Eq. (2.3) where the probabilities and consequences of all potentially dangerous events are combined together in a single risk value.

From the above said, an informative and operative definition of risk should allow answering the three fundamental questions of any risk analysis [1], [2], [5], [7], [8], [9], [10]:

– Which sequences of undesirable events transform the hazard into an actual damage?
– What is the probability of each of these sequences?
– What are the consequences of each of these sequences?

The answers to these questions lead to a definition of risk in terms of a set of triplets [5]:

$$R = \left\{ \left\langle s_i, p_i, x_i \right\rangle \right\}$$

where s_i is the sequence of undesirable events leading to damage, p_i is the associated probability and x_i the consequence. Thus, the outcome of a risk analysis is a list of scenarios, such as the one in Table 2.1, which represents the risk.

Table 2.1: Risk as a list of triplets

Sequence	Probability	Consequence
s_1	p_1	x_1
s_2	p_2	x_2
...	...	...
s_n	p_n	x_n

On the basis of this information, the designer, the manager and the regulator, can act effectively so as to reduce risk.

2.3   Risk analysis

From the previous definition of risk, it is evident that a rational management of it entails a proper treatment of the uncertainties associated with the occurrence of accidental scenarios.

Classically, the management and control of the risk associated to a given plant has been based on the definition of a group of sequences of events leading to undesired consequences, representing credible *worst-case* accident scenarios, $\{ s^* \}$, and on the prediction and analysis of their consequences, $\{ x^* \}$. Then, the safety and protection of the system is designed against such events (*design-basis accidents*), to prevent them and to protect from, and mitigate, their associated consequences.

This *structuralist defense-in-depth* viewpoint and the safety margins derived from it, have been embedded into conservative regulations under the creed that the identified worst-case, credible accidents would envelope all credible accidents, for what concerns the challenges and stresses posed on the system and its protections. The underlying principle has been that if a plant were designed to withstand all the large credible accidents, then it would be 'by definition' protected against any credible accident.

This approach has been the one classically undertaken, and in many instances it still is, to protect a plant from the uncertainty of the unknown failure behaviours of its components, systems and structures, without quantifying it, and to provide reasonable assurance that a plant can be operated without undue risk.

However, the practice of referring to "worst" cases implies a high level of subjectivity and arbitrariness which may lead to the consideration of scenarios characterized by really catastrophic consequences, albeit highly unlikely. This somewhat arbitrary approach to safety can lead to excessive conservatism, with a penalization of the industry due to the imposition of unnecessarily stringent regulatory burdens. This is particularly so for those industries, such as the nuclear one, in which accidents may lead to potentially large consequences.

With the growing use of the nuclear energy in the 1960s, the need soon arose for a more rational and logical approach to the design, regulation, operation and management of hazardous systems. A new viewpoint was then proposed, based on the analysis of the reliability of the consequence-limiting protection systems involved in all potential accident scenarios, with no longer any differentiation between credible and incredible, large and small [3].

The nuclear community in Canada, in particular, was a strong supporter of such a probabilistic approach to safety. This was mainly due to the consideration that their nuclear reactor design, the Pressurized Heavy Water Reactor (PHWR) is indeed characterized by an intrinsically dangerous physical feature (the so called *positive temperature feedback* which could lead to a dangerous escalation of the nuclear reaction process, under certain conditions), but on the other hand this unsafe feature is counteracted by a highly-reliable and quick shut-down system (which works at low pressure and temperature and is separated by the primary cooling circuit whose failure cannot damage it). Then, the only way of accounting for this aspect is the introduction in the analysis of probabilistic measures of the likelihood of the accident scenarios.

This sparked a series of studies in the mid 1960s aimed at investigating the merits of a more quantitative approach, based on probability, to the treatment of uncertainty associated with accident scenarios [4]. The findings of these studies motivated the first complete and full-scale probabilistic risk assessment of a nuclear power installation [10]. This extensive work showed that indeed the dominant contributors to risk need not be necessarily the design-basis accidents, a 'revolutionary' discovery undermining the fundamental creed underpinning the structuralist, defense-in-depth approach to safety.

Along these lines of thoughts and after several 'battles' for demonstration and recognition, a new approach to risk analysis has arisen, not limited only to the consideration of worst-case accident scenarios but which looks to all feasible scenarios and its related consequences, with the probability of occurrence of such scenarios becoming an additional key aspect to be quantified in order to rationally handle uncertainty [1], [2], [5], [7], [8], [9], [10].

On this basis, new regulatory criteria have been introduced, which account for both the consequences of the scenarios and their probabilities of occurrence under a now *rationalist* defense-in-depth approach. An example of criterion of this kind can be represented graphically as shown in Figure 2.2, where the probabilities p of the scenarios are plotted against their consequences x (the so-called *Farmer curve*). A proportionality line divides the (x, p) space into two zones: scenarios above such line (i.e. in the dark zone) lead to unacceptable risks whereas those below (i.e. in the clear zone) represent acceptable risks. This allows accepting risks associated to scenarios characterized by high consequences, provided they have very low probability of occurrence (e.g. the point in Figure 2.2).

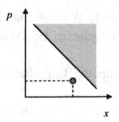

Fig. 2.2: Probability-consequence curve

When, as depicted in Figure 2.2, the slope is −1, probability and consequence carry the same importance in defining the risk level; on the contrary, when more emphasis is placed on consequences than on probabilities, the slope of the line is increased, as shown in Figure 2.3.

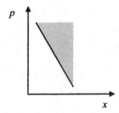

Fig. 2.3: Probability – consequence curve with a slope such that consequences have more importance than probability

The introduction of probabilities in the consideration of risk has been very controversial from the beginning and only after several years the proper recognition has been given to its usefulness for a balanced evaluation and a rational management of risk.

For further insights in the Subject, the interested reader is advised to consult the specialized technical literature, e.g. [1], [2], [5], [7], [8], [9].

References

[1]   Aven, T., Foundations of Risk Analysis, Wiley, 2003.
[2]   Bedford, T. and Cooke, R., Probabilistic Risk Analysis, Cambridge University Press, 2001.
[3]   Farmer, F.R., The Growth of Reactor Safety Criteria in the United Kingdom, Anglo-Spanish Power Symposium, Madrid, 1964.

[4]   Garrick, B.J. and Gekler, W.C., Reliability Analysis of Nuclear Power Plant Protective Systems, US Atomic Energy Commission, HN-190, 1967.

[5]   Henley, E.J. and Kumamoto, H., Probabilistic risk assessment, NY, IEEE Press, 1992.

[6]   Kaplan, S. and Garrick, B. J., Risk Analysis, 1, p. 1-11, 1984.

[7]   McCormick, N.J., Reliability and risk analysis, New York, Academic Press, 1981.

[8]   PRA Procedures Guide, Vols. 1&2, NUREG/CR-2300, January 1983.

[9]   Probabilistic Risk Assessment Procedures Guide for NASA Managers and Practitioners, NASA, 2002.

[10] WASH-1400, Reactor Safety Study, US Nuclear Regulatory Commission, 1975.

3

Methods for Hazard Identification

The first step into the analysis of the risk of a given system is that of identifying the hazards associated to its operation. The output of this task consists of a list of the sources of potential danger, i.e. those accident initiators (component failures, process deviations, external events, operator errors) which have a probability of occurrence not equal to zero and which can give rise to significant consequences. The identification of the accident initiators is obviously a key aspect of the overall safety analysis and great care must be put into its completeness since those accident events not included at this stage are very unlikely to enter in the analysis at a later stage.

The methods developed for performing this step consist, in general, in a qualitative analysis of the system and its functions, within a systematic framework of procedures. The methods strongly rely on the expertise of the designers, analysts and personnel who have designed, operated and maintained the system. Some of the methodologies most commonly used are:

1. *Check list*
2. *Hazard index method*
3. *Hierarchical trees*
4. *System Identification of Release Points (SIRP)*
5. *Failure Mode and Effect Analysis (FMEA)*
6. *HAZard and OPerability analysis (HAZOP)*

Such methodologies are not mutually exclusive but, rather, they are often used in a complementary way.

As the first two methods are of straightforward application, here we limit ourselves to giving few insights into the principles of the other four methods in the list. For more details, the interested reader should consult the specialized literature, e.g. [1].

3.1   Hierarchical trees

This deductive method allows the identification of the initiating causes of a pre-specified, undesired event, through the development of a structured logic tree. Obviously, such event must be known a priori. That is the reason why it was initially developed for the nuclear industry, where the undesired event (e.g. offsite release of radioactive material) is well defined a priori.

The construction of the tree (see Fig. 3.1 for an example related to the hazard of an offsite release from a nuclear power plant) starts at the top with the undesired event (*offsite release*) at the *public impact* level; the undesired event may occur due to various pathways (*release of core/non-core material*) which are explicited as independent branches at the *damage pathway* level; these pathways are generated due to loss of the various containments and/or mitigation functions which are indicated in the tree at the *containment* or *mitigation* level; the containment function becomes necessary after loss of the devised safety functions which should prevent the accident and which are listed at the *safety function* level so that to each containment and mitigation barrier we can associate the correlated safety functions; finally, such safety functions can be linked to the primary initiator events which require such safety functions and which constitute the root causes (*initiating event* level) of the top event of the tree.

Note that this approach finds its natural application in systems, such as the nuclear and aerospace ones, which have been designed in a safety-oriented manner so that safety functions, mitigation and containment barriers etc. are clearly and uniquely defined.

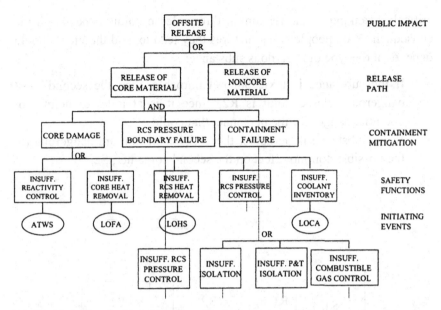

Fig. 3.1: Example of a hierarchical tree for the hazard of offsite release from a nuclear power plant

3.2 Systematic Identification of Release Points (SIRP)

This approach aims at identifying the points of most likely release on the basis of historical data. Ducts, containments, release valves and rupture discs are identified as potential release points. Given the design and structural characteristics of these items, historical data is used to associate to them a most likely dimension of break and a probability of break occurrence. Expert judgment is then used to eliminate those break points whose position, dimension and probability are such to render the consequences irrelevant. Finally, *equivalent, reference* break points are identified in the circuit for grouping those break points leading to similar accident evolutions.

Figure 3.2 reports an example in which four release points, R1, R2, R3, R4, have been identified. R1 is obviously the most critical point since a break in this position would lead to the release of the whole amount of hazardous substance contained in the tank S1, whereas the other break points can be isolated by proper action on the valve V. The criticality ranking of the remaining three release points depends on the consequences on people and structures they lead to, and therefore should depend on the type of hazardous substance:

–   If the substance is toxic but not burnable, then the second most dangerous release point is R2, since it is at a lower height, of potential danger for the personnel that could be around;
–   If the substance is burnable, then R3 becomes more dangerous for the possible domino effect on the second container, S2.

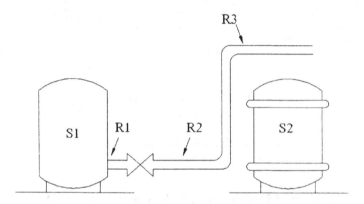

Fig. 3.2: Example of a system and its identified points of release

3.3   Failure Mode and Effect Analysis (FMEA)

This is a qualitative method, of inductive nature, which aims at identifying those failure modes of the components which could disable system operation or become initiators of accidents with significant external consequences.

The analysis proceeds as follows:

1.   Decompose the system in functionally independent subsystems; for each subsystem identify the various operation modes (start-up, regime, shut-down, maintenance, etc.) and its configurations when operating in such modes (valves open or closed, pumps on or off, etc.).

2.   For each subsystem in each of its operation modes, compile a Table such as Table 3.1, without neglecting any of the subsystem components. The Table for a component should include its failure modes and the effects that such failure has on other components, on the subsystem and on the whole plant.

The analysis considers only the effects of single-failures, except for the case of stand-by components for which the effects of its failure are obviously considered only conditioned on the demand of its intervention due to the failure of the main component. In general, then, there is no indication of the risk associated with multiple or common cause failures.

To ensure a coherent analysis, the analyst must be sure that similar components are given the same failure modes, with same probability qualifications.

An extension of FMEA often employed in practice is FMECA (Failure Mode, Effect and Criticality Analysis) in which a criticality class is assigned to each failure mode according to the following ranking:

Safe   =   no relevant effects;
Marginal   =   partially degraded system but no damage to humans;
Critical   =   system damaged and damages also to humans; if no protective actions are undertaken the accident could lead to loss of the system and serious consequences on the humans;
Catastrophic = Loss of the system and serious consequences on humans.

Figure 3.3 presents a simple system whose FMECA Table is reported in Table 3.2.

If the analysis is carried out in the design phase, it is difficult to base the analysis on the components (yet to be defined) and on their failure modes; in this case, the analysis can refer to the different functions required to the subsystems and the effects of the functions not being performed.

For complex systems, a FMEA can be rather burdensome. There are several computer tools available on the market which guide the implementation of such techniques, with friendly check lists.

Table 3.1: Typical FMECA Table [1]

SYSTEM:									
OPERATION MODE:									
Component	Failure mode	Effects on other components	Effects on subsystem	Effects on plant	Probability	Criticality	Detection methods	Protections and mitigation	Remarks
Description	Failure modes relevant for the operational mode indicated	Effects of failure mode on adjacent components and surrounding environment	Effects on the functionality of the subsystem	Effects on the functionality and availability of the entire plant	Probability of failure occurrence (usually qualitative)	Criticality rank of the failure mode on the basis of its effects and probability (qualitative estimation of risk)	Methods of detection of the occurrence of the failure event	Protections and measures to avoid the failure occurrence	Remarks and suggestions on the need to consider the failure mode as accident initiator

Overall, the procedure of FMEA is rather simple and schematic, and it allows one to carefully analyze the whole system. Often, this analysis is used in support of the construction of fault trees (Chapter 7) and of reliability-centered maintenance programs to find optimal maintenance strategies. In the latter case, the effects and criticality of the various failure modes are examined not only from the safety viewpoint but also from that of plant availability.

3.3.1 Example of FMECA: Domestic Hot Water System

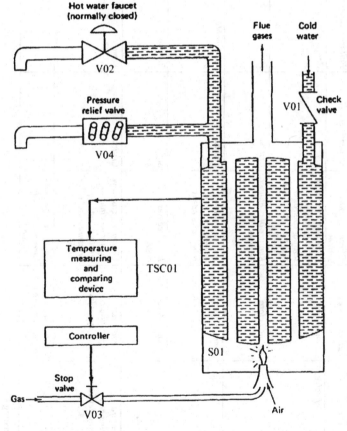

Fig. 3.3: Schematic of the domestic hot water system [1]

Table 3.2: FMECA Table for the domestic hot water system [1]

Component	Failure mode	Effects on whole system	Critically class	Failure frequency	Detection methods	Compensating provision and remarks
Pressure relief valve (V04)	Jammed open	Increasing operation of temperature sensing controller; Gas flow due to hot water loss	Safe	Reasonably probable	Observe at pressure relief valve	Shut off water supply, reseal or replace relief valve
	Jammed close	Rupture of container or pipes	Critical	Probable	Manual testing	If combined with other component failures, otherwise this failure has no consequence
Gas valve (V03)	Jammed open	Burner continues to operate, pressure relief valve opens	Critical	Reasonably probable	Water at faucet too hot: pressure relief valve open (observation)	Open hot water faucet to relieve pressure. Shut off gas supply. Pressure relief valve compensates. IE1.
	Jammed close	Burner ceases to operate	Safe	Remote	Observe at output (Water temperature too low)	
Temperature measuring and comparing device (Tsc01)	Fail to react to temperature rise above preset level	Controller, gas valve, burner continue to function "on". Pressure relief valve opens	Critical	Remote	Observe at output (faucet)	Pressure relief valve compensates. Open hot water faucet to relieve pressure. Shut off gas supply. IE2.
	Fail to react to temperature drop below preset level	Controller, gas valve, burner continue to function "off".	Safe	Remote	Observe at output (faucet)	

IE: initiating Event

3.4   HAZard and OPerability analysis (HAZOP)

HAZOP is a qualitative methodology which combines deductive aspects (search for causes) and inductive aspects (consequence analysis) with the objective of identifying the initiating events of undesired accident sequences. Contrary to FMEA, which is mainly based on the structural/hardware aspects of the system, HAZOP looks at the processes which are undergoing in the plant. Indeed, the method, initially developed for the chemical process industry, proceeds through the compilation of Tables (such as Table 3.3) which highlight possible process anomalies and their associated causes and consequences.

The analysis proceeds as follows:

1.   Decompose the system in functionally independent process units (reaction unit, storage unit, pumping unit, etc.); for each process unit identify the various operation modes (start-up, regime, shut-down, maintenance, etc.).

2.   For each process unit and operation mode, identify the potential deviations from the nominal process behaviour. In order to do this, we must:

   a)   specify all the unit incoming and outgoing fluxes (energy, mass, control signals, etc.) and the characteristic process variables (temperature, flow rate, pressure, concentration, etc.);

   b)   write down the various functions that the unit is supposed to attend (heating, cooling, pumping, filtering, etc.).

   c)   apply keywords such as *low, high, no, reverse*, etc., to the previously identified process variables and unit functions, so as to generate deviations from the nominal process regime.

3.   For each process deviation, qualitatively identify its possible causes and consequences. For the consequences, include effects also on other units: this allows HAZOP to account also for domino effects among different units.

On the market, there are software tools available to guide an HAZOP analysis.

Table 3.3: Typical HAZOP Table [1]

UNIT:					
OPERATION MODE:					
Keyword	**Deviation**	**Cause**	**Consequence**	**Hazard**	**Actions needed**
More	More Temperature	Additional Thermal resistance	Higher pressure in tank	Release due to Overpressure	Install high temperature warning and pressure relief valve

References

[1] Henley, E.J. and Kumamoto, H., Probabilistic risk assessment, NY, IEEE Press, 1992.

4

Basics of Probability Theory for Applications to Reliability and Risk Analysis

4.1 Definitions

In probabilistic terminology an *experiment* ε is defined as a process whose outcome is a priori unknown to the analyst. The possible outcomes are all a priori known and classified but which one will occur is unknown at the time the experiment is performed. This definition is consistent with the Bayesian view of probability according to which the outcome of an experiment may be deterministic, but at the moment unknown (e.g. the current number of sons of a friend with which one has lost contact long time ago) as well as stochastic (e.g. the result of a dice toss).

To each experiment ε is associated a *sample space* Ω, which represents the set of all possible outcomes of ε. The sample space can be discrete finite (e.g. for an experiment of a coin or dice toss), countably infinite (e.g. the number of persons crossing the street in a given period of time: in principle, it could be infinite and yet be counted) or continuous (e.g. the value of the dollar currency in the year 3012).

An *event* E is then a group of possible outcomes of the experiment ε, i.e. a subset of Ω. In particular, each possible outcome represents an (*elementary*) *event* itself, being a subset of Ω. Further, the *null set* $\varnothing$ and the sample space Ω can also be considered events.

To each event E is possible to associate its complementary event $\overline{E}$, constituted by all possible outcomes in Ω which do not belong to E.

We say that event E occurs when the outcome of the experiment ε is one of the elements of E.

4.2   Boolean logic operations

In the *logic of certainty* (Boolean logic), an event can either occur or not occur. Thus, it is represented by a statement, or proposition which can only be either *true* or *false*, and at a certain point in time, after the experiment is performed, the analyst will know its actual state.

Correspondingly, to event E we can associate an indicator variable X_E which takes the value of 1 when the event occurs in the experiment and 0 when it does not. As a counter-example, the statement "It may rain tomorrow" does not represent an event because it does not imply a "true" or "false" answer. We define the following operations involving Boolean events:

Negation: Given event E, represented by the indicator variable X_E, its negation $\overline{E}$ is described by

$$\overline{X}_E = 1 - X_E \qquad (4.1)$$

Union: The event $A \bigcup B$, union of the two events A and B, is true, e.g. $X_{A \cup B} = 1$, if any one of A or B is true. Hence,

$$X_{A \cup B} = 1 - (1 - X_A)(1 - X_B)$$
$$= 1 - \prod_{j=A,B} (1 - X_j) = \coprod_{j=A,B} X_j = X_A + X_B - X_A X_B \qquad (4.2)$$

Often in practice this event is indicated as $A+B$.

Intersection: The event $A \bigcap B$, intersection of the events A and B, is true, e.g. $X_{A \cap B} = 1$, if both A and B are simultaneously true. Hence,

$$X_{A \cap B} = X_A X_B \qquad (4.3)$$

Often in practice this event is indicated as AB and referred to as the joint event A and B.

Mutually exclusive events: Two events A and B are said to be mutually exclusive if their intersection is the null set, i.e.

$$X_{A \cap B} = 0 \qquad (4.4)$$

Example 4.1 [1]

Strong wind at a particular site may come from any direction between due east ($\theta = 0°$) and due north ($\theta = 90°$). All values of wind speed V are possible.

1. Sketch the sample space for wind speed and direction.
2. Let $A = \{V > 20 \text{ mph}\}$
    $\quad B = \{12 \text{ mph} < V \le 30 \text{ mph}\}$
    $\quad C = \{\theta \le 30°\}$
    Identify the events A, B, C, and $\overline{A}$ in the sample space sketched in part 1.
3. Use new sketches to identify the following events:

    (i) $D = A \cap C$

    (ii) $E = A \cup B$

    (iii) $F = A \cap B \cap C$

4. Are the events D and E mutually exclusive? How about events A and C?

Solution

4.1.1 Sample Space for wind speed and direction

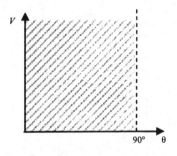

Fig. 4.1: Shaded area represents Sample Space

4.1.2 Sketches of Events

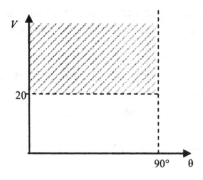

Fig. 4.2: Shaded area represents Event A = {V > 20 mph}

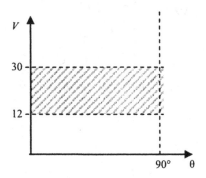

Fig. 4.3: Shaded area represents Event B = {12 < V ≤ 30}

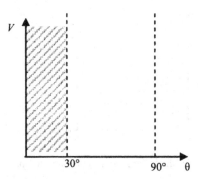

Fig. 4.4: Shaded area represents Event C = {θ ≤ 30⁰}

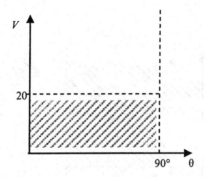

Fig. 4.5: Shaded area represents Event $\overline{A}$ = {V ≤ 20mph}

4.1.3 Sketches of Events

(i) $D = A \cap C$

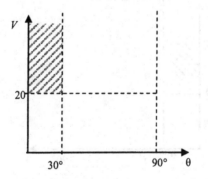

Fig. 4.6: Shaded area represents Event D = A ∩ C

(ii) $E = A \bigcup B$

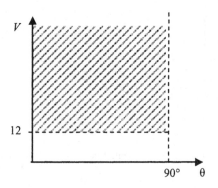

Fig. 4.7: Shaded area represents Event $E = A \cup B$

(iii) $F = A \cap B \cap C$

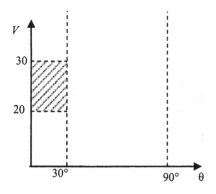

Fig. 4.8: Shaded area represents Event $F = A \cap B \cap C$

4.1.4 Mutually Exclusive

D and E are not mutually exclusive. (Because $D \cap E \neq \varnothing$, in fact $D \cap E = D,$).

A and C are not mutually exclusive. (Because $A \cap C \neq \varnothing$, in fact $A \cap C = D$).

4.3 Logic of uncertainty: definition of probability

As previously explained, for a statement to be an event, it can only have two possible states, either true or false, and at a certain point in time the exact state will become known as a result of the actual performing of the associated experiment.

4.3.1 Axiomatic Definition

At the current state of knowledge it is possible that the state of an event be *uncertain*, although at some point in the future uncertainty will be removed and replaced by either the true or the false state. Inevitably, if one needs to make decisions based on the current state of knowledge, he has to deal with such uncertainty. In particular, one needs to be able to compare different uncertain events and say whether one is more likely to occur than another. Hence, we accept the following *axiom* as a primitive concept which does not need to be proven:

Uncertain events can be compared

It represents a concept very similar to that of the value of objects and goods which need to be compared for the purpose of exchanging them. In this latter case at one point in history, the monetary scale was introduced as an absolute scale against which to compare different goods with respect to their values. Similarly, it is necessary to introduce a measure for comparing uncertain events.

Let us consider an experiment ε and let Ω be its sample space. To each event E we assign a real number $p(E)$, which we call *probability of E* and which satisfies the following three Kolmogorov axioms:

I.   For each event E, $0 \le p(E) \le 1$

II. For event Ω, it is $p(\Omega) = 1$; for event $\varnothing$, it is $p(\varnothing)=0$.

III. Let $E_1, E_2,..., E_n$ be a finite set of mutually exclusive events. Then,

$$p\left(\bigcup_{i=1}^{n} E_i\right) = \sum_{i=1}^{n} p(E_i)$$

(4.5)

The latter axiom is called the *addition law* and is assumed to maintain its validity also in the case of countably infinite sample spaces.

This axiomatic view constitutes the Bayesian, or subjectivist, interpretation of probability according to which everything is made relative to an assessor which declares 'a priori' its 'belief' regarding the likelihood of uncertain events in order to be able to compare them. Thus, in this view, the probability of an event E represents a *degree of belief*, or *degree of confidence*, of the assessor with regards to the occurrence of that event. In other words, probability is nothing more than a measure of uncertainty about the likelihood of an event. A probability assignment is a numerical encoding of a state of knowledge of the assessor, rather than a property of the 'real world'. Because the probability assignment is subjectively based on the assessor's internal state, in most practical situations there is no 'true' or 'correct' probability for a given event and the probability value can change as the assessor gains additional information (experimental evidence). Obviously, it is completely 'objective' in the sense that it is independent of the personality of the assessor who must assign probabilities in a coherent manner, which requires obeying to the axioms and laws of probability, in particular to Bayes theorem for updating the probability assignment on the basis of newly collected evidence (see Section 4.4.4 below). By so doing, two assessors sharing the same total background of knowledge and experimental evidence on a given event must assign the same probability for its occurrence.

4.3.2   Empirical Frequentist Definition

Let E be an event associated to experiment ε. Suppose that we repeat the experiment n times and let k be the number of times that event E occurs. The ratio k/n represents the relative frequency of occurrence of E. As the number of repetitions n approaches infinity we empirically observe that the ratio k/n settles around an asymptotic value, p, and we say that p is the probability of E.

From a rigorous point of view, this empirical procedure does not follow the usual definition of mathematical limit and it can be synthesized as follows:

$$\lim_{n\to\infty}\left|\frac{k}{n}-p\right|<\xi \qquad (4.6)$$

with $\xi > 0$. Obviously, this definition may be somewhat unsatisfactory as probability is defined in terms of likelihood of a large number of repeated experiments.

4.3.3 Classical Definition

This definition is very similar to the empirical one of the previous Section 4.3.2. The only fundamental difference is that it is not necessary to resort to the procedure of taking a limit. Let us consider an experiment with N possible elementary, mutually exclusive and equally probable outcomes $A_1, A_2,..., A_N$. We are interested in the event E which occurs if anyone of M elementary outcomes occurs, $A_1, A_2,..., A_M$, i.e. $E = A_1 \cup A_2 \cup ... \cup A_M$.

Since the events are mutually exclusive and equally probable,

$$p(E)=\frac{number\ of\ outcomes\ of\ interest}{total\ number\ of\ possible\ outcomes} \qquad (4.7)$$

This result is very important because it allows computing the probability with the methods of combinatorial calculus; its applicability is however limited to the case in which the event of interest can be decomposed in a finite number of mutually exclusive and equally probable outcomes. Furthermore, the classical definition of probability entails the possibility of performing repeated trials; it requires that the number of outcomes be finite and that they be equally probable, i.e. it defines probability resorting to a concept of frequency.

4.3.4   Probability space

Once a probability measure is defined in one of the above illustrated ways, the mathematical theory of probability is founded on the three fundamental axioms of Kolmogorov introduced in Section 4.3.1, independently of the definition. All the theorems of probability follow from these three axioms.

When assigning probability values to events of a sample space, a difficulty arises for continuous sample spaces, e.g. $\Omega \equiv (0,1)$. Indeed, continuous intervals cannot be constructed by adding elementary points in a countable manner and correspondingly, probabilities of continuous intervals cannot be assigned by the addition law of probability. In other words, if we were to assign to each $E \in (0,1)$ a probability $p(E)$, then the sum of all $p(E)$'s would go to infinity, unless $p(E) = 0$ for 'almost all' $E \in (0,1)$.

The way to overcome this difficulty is to assign a probability not to each individual outcome E but to subsets of Ω. For example, one could define the probability of a subset $A \equiv (a,b) \subset (0,1)$ as the measure $l(A) = b - a$. In particular, each countable set of individual outcomes $\{E_k\}$, taken as the interval (E_k, E_k), has null measure and, thus, zero probability. By so doing, it is possible to assign a measure, and thus a probability to any set A made of unions, intersections and complements of intervals. Still, there are *ill* sets which cannot be constructed as explained, to which it is not possible to assign probabilities coherently with the third Kolmogorov axiom and which are, thus, termed *not probabilizable*. From the theoretical viewpoint, this difficulty is overcome by limiting our consideration to one of the many families F of subsets of Ω which are *well-behaved*. Such family is called a σ-*algebra* and we assign probability values only to subsets belonging to F: correspondingly, the term *event* refers only to such subsets. In more details, a σ-*algebra* is a family F of subsets of Ω which satisfies the following conditions:

i.   If $E \in F$ then also $\overline{E} = \Omega - E \in F$

ii. If $E_1, E_2, \ldots$ is a countable infinity of subsets in F, then $\bigcup\limits_{i=1}^{\infty} E_i \in F$

and $\bigcap\limits_{i=1}^{\infty} E_i \in F$.

In words, a *σ-algebra* is a family of sets of the space Ω which is closed with respect to the operation of complement and to the formation of a countable infinity of unions and intersections.

Since the space Ω is the union of E and $\overline{E}$, it belongs to the *σ-algebra*, i.e. $\Omega \in F$. Examples of *σ-algebra* are:

- The *largest* *σ-algebra* in Ω is the family of all subsets of Ω.
- The *smallest* *σ-algebra* in Ω consists of Ω and the null set $\varnothing$.
- Let us consider the space $\Omega \equiv \Re^1$ and a *σ-algebra* F constituted by subsets of E. If to each $x \in E$ we associate all values $x \pm 1, x \pm 2, \ldots$ we obtain another *σ-algebra*.

The triplet (Ω, F, p) defines the *probability space*.

4.4 Probability laws

As previously mentioned, to the generic random event E is associated an indicator variable X_E which takes the value of 1 when the event occurs in the experiment and 0 when it does not. Correspondingly, a real number $p(E)$ is assigned to measure the *probability of* E and which satisfies the three Kolmogorov axioms. Given the binary nature of the indicator variable, X_E can only take values of 0 or 1 so that:

$$p(E) = p(X_E = 1) \cdot 1 + p(X_E = 0) \cdot 0 = E[X_E] \qquad (4.8)$$

4.4.1 Union of non-mutually exclusive events

Consider n events E_n not mutually exclusive. Their union E_U is associated with an indicator variable X_U which is the extension of the formula (4.2) for the union of the two events A and B. For example, for the intersection of the three events A, B and C we have

$$X_U = 1 - \prod_{j=A,B,C} (1 - X_j) = 1 - (1 - X_A)(1 - X_B)(1 - X_C) =$$

$$= X_A + X_B + X_C - X_A X_B - X_A X_C \qquad (4.9)$$
$$- X_B X_C + X_A X_B X_C$$

Following (4.8), the probability of the event E_U can then be computed applying to (4.9) the (linear) expectation operator. More generally, for the union of n non-mutually exclusive events:

$$P(E_U) = E[X_U] = \sum_{j=1}^{n} E[X_j] - E[\sum_{i=1}^{n-1} \sum_{j=i+1}^{n} X_i X_j] + \ldots + (-1)^{n+1} \prod_{j=1}^{n} E[X_j] =$$

$$= \sum_{j=1}^{n} P(E_j) - \sum_{i=1}^{n-1} \sum_{j=i+1}^{n} P(E_i \cap E_j) + \ldots + (-1)^{n+1} \prod_{j=1}^{n} P(E_j)$$

$$(4.10)$$

From an engineering practice point of view, it is often necessary to introduce reasonably bounded approximations of (4.10). Keeping only the first sum, one obtains an upper bound,

$$P(E_U) \le \sum_{j=1}^{n} P(E_j) \qquad (4.11)$$

whereas keeping the first two sums gives a lower bound,

$$P(E_U) \ge \sum_{j=1}^{n} P(E_j) - \sum_{i=1}^{n-1} \sum_{j=i+1}^{n} P(E_i \cap E_j) \qquad (4.12)$$

More refined upper and lower bounds can then be obtained by alternately keeping an odd or even number of sum terms in (4.10).

Since in reliability and risk calculations the probability of high-order joint events is very small , it is common practice to use the upper bound (4.11), which is often referred to as the *rare-event* approximation.

4.4.2   Conditional Probability

In many practical situations, it is important to compute the probability of an event A given that another event B has occurred. This probability is called the *conditional probability of A given B* and it is given by the ratio of the probability of the joint event $A \cap B$ over the probability of the conditioning event B, viz.

$$P(A \mid B) = \frac{P(A \cap B)}{P(B)} \qquad (4.13)$$

Intuitively, $P(A \mid B)$ gives the probability of the event A not on the entire possible sample space Ω but on the sample space relative to the occurrences of B. This is the reason for the normalization by $P(B)$ of the probability of the joint event $P(A \cap B)$ in (4.13).

Based on the conditional probability, it is possible to introduce the concept of statistical independence: event A is said to be statistically independent from event B if $P(A \mid B) = P(A)$. In other words, knowing that B has occurred does not change the probability of A. From (4.13), it follows that if A and B are statistically independent $P(A \cap B) = P(A)P(B)$. Note that the concept of statistical independence should not be confused with that of mutual exclusivity ($X_A X_B = 0$, Section 4.2) which is actually a logical dependence: knowing that A has occurred ($X_A = 1$), guarantees that B cannot occur ($X_B = 0$).

Example 4.2 [1]

There are two streams flowing past an industrial plant. The dissolved oxygen, DO, level in the water downstream is an indication of the degree of pollution caused by the waste dumped from the industrial plant. Let A denote the event that stream a is polluted, and B the event that stream b is polluted. From measurements taken on the DO level of each stream over the last year, it was determined that in a given day

$$P(A) = 2/5 \quad and \quad P(B) = 3/4$$

and the probability that at least one stream will be polluted in any given day is $P(A \cup B) = 4/5$.

1.   Determine the probability that stream a is also polluted given that stream b is polluted.
2.   Determine the probability that stream b is also polluted given that stream a is polluted.

Solution

First, we compute the probability that both streams are polluted. Since

$$P(A \cup B) = P(A) + P(B) - P(A \cap B)$$

We have

$$
\begin{aligned}
P(A \cap B) &= P(A) + P(B) - P(A \cup B) \\
&= (2/5) + (3/4) - (4/5) \\
&= (7/20)
\end{aligned}
$$

4.2.1 $P(A|B)$

$$P(A|B) = \frac{P(A \cap B)}{P(B)} = \frac{7/20}{3/4} = 7/15 = 0.46$$

4.2.2 $P(B|A)$

$$P(B|A) = \frac{P(A \cap B)}{P(A)} = \frac{7/20}{2/5} = 7/8 = 0.875$$

In other words, stream b is very likely to be polluted when stream a is polluted, whereas chances are less than 50% that stream a will be polluted when stream b is polluted.

4.4.3 Theorem of Total Probability

Let us consider a partition of the sample space Ω into n mutually exclusive and exhaustive events E_j, $j = 1, 2, ..., n$. In terms of Boolean events, this is written as:

$$E_i \cap E_j = 0 \;\; \forall i \neq j \qquad \bigcup_{j=1}^{n} E_j = \Omega \qquad (4.14)$$

whereas in terms of the indicator variables,

$$X_i X_j = 0 \;\; \forall i \neq j \qquad \sum_{j=1}^{n} X_j = 1 \qquad (4.15)$$

Given any event A in Ω, its probability can be computed in terms of the partitioning events E_j, $j = 1, 2, ..., n$ and the conditional probabilities of A on these events, viz.

$$P(A) = P(A | E_1)P(E_1) + P(A | E_2)P(E_2) + ... + P(A | E_n)P(E_n) \quad (4.16)$$

Example 4.3 [1]

The air pollution in a city is caused mainly by industrial *(I)* and automobile *(A)* exhausts. In the next 5 years, the chances of successfully controlling these two sources of pollution are, respectively, 75% and 60%. Assume that if only one of the two sources is successfully controlled, the probability of bringing the pollution below acceptable level would be 80%.

1. What is the probability of successfully controlling air pollution in the next 5 years?
2. If, in the next 5 years, the pollution level is not sufficiently controlled, what is the probability that is entirely caused by the failure to control automobile exhaust?
3. If pollution is not controlled, what is the probability that control of automobile exhaust was not successful?

Solution

A = event of successful control of the automobile exhausts
I = event of successful control of the industrial exhausts
E = event of bringing the pollution below the acceptable level

From the problem statement we have:

$P(I) = 0.75$

$P(A) = 0.60$

and

$P(E|\overline{A}I) = P(E|A\overline{I}) = 0.8$
$P(E|\overline{A}\overline{I}) = 0$
$P(E|AI) = 1$

4.3.1 Probability of controlling air pollution in the next 5 years

The possible combinations of the two pollution events are $AI, A\overline{I}, \overline{A}I, \overline{A}\overline{I}$. If we assume statistical independence between controlling industrial (I) and automobile (A) exhausts, we have:

$$P(AI) = 0.60 \cdot 0.75 = 0.45$$
$$P(A\overline{I}) = 0.60 \cdot 0.25 = 0.15$$
$$P(\overline{A}I) = 0.40 \cdot 0.75 = 0.30$$
$$P(\overline{A}\overline{I}) = 0.40 \cdot 0.25 = 0.10$$

$AI, A\overline{I}, \overline{A}I, \overline{A}\overline{I}$ are mutually exclusive and collectively exhaustive events. Then, we can use the theorem of total probability (Fig. 4.9):

$P(E) = P(E|AI)\ P(AI)$
$\qquad + P(E|A\overline{I})\ P(A\overline{I}) + P(E|\overline{A}I)\ P(\overline{A}I) + P(E|\overline{A}\overline{I})\ P(\overline{A}\overline{I}) = 0.81$

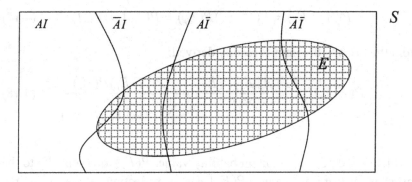

Fig. 4.9: Partitioning of event E

4.3.2 Probability of pollution not controlled due to failure of controlling automobile exhaust

$$P(\overline{AI} \mid \overline{E}) = \frac{P(\overline{E} \mid \overline{AI})P(\overline{AI})}{P(\overline{E})} = \frac{[1 - P(E \mid \overline{AI})]P(\overline{AI})}{P(\overline{E})} = 0.32$$

4.3.3 Probability of automobile exhaust not controlled given that pollution is not controlled

$$P(\overline{A} \mid \overline{B}) = P(\overline{AI} \cup \overline{A\overline{I}} \mid \overline{E}) = P(\overline{AI} \mid \overline{E}) + P(\overline{A\overline{I}} \mid E) = \frac{P(\overline{E} \mid \overline{AI})P(\overline{AI})}{P(\overline{E})} + \frac{P(\overline{E} \mid \overline{A\overline{I}})P(\overline{A\overline{I}})}{P(\overline{E})}$$

$$= \frac{0.2 \cdot 0.3}{0.19} + \frac{1 \cdot 0.1}{0.19} = 0.84$$

4.4.4 Bayes theorem

Assume now that you have experimental evidence that event A has occurred. What is the probability that event E_i has also occurred? This may be considered as a 'reverse' probability with respect to the probability question underlying the previous theorem of total probability. To the joint event $A \cap E_i$ we can apply the conditional probability (4.13) from both the points of view of A and of E_i,

$$P(A \cap E_i) = P(A \mid E_i)P(E_i) = P(E_i \mid A)P(A) \qquad (4.17)$$

From this, Bayes theorem is readily derived:

$$P(E_i \mid A) = \frac{P(A \mid E_i)P(E_i)}{P(A)} = \frac{P(A \mid E_i)P(E_i)}{\displaystyle\sum_{j=1}^{n} P(A \mid E_j)P(E_j)} \qquad (4.18)$$

Eq. (4.18) updates the *prior* probability value $P(E_i)$ of event E_i to the *posterior* probability value $P(E_i \mid A)$ in reflection of the acquired experimental evidence on the occurrence of event A whose unknown probability $P(A)$ is computed by applying the theorem of total probability (4.16).

Thus, coherently with the Bayesian definition of probability, the assignment of the probability measure of an event depends on the knowledge that the assessor has relative to such event. If such state of knowledge changes, then the probability assignment must change accordingly, coherently with the Kolmogorov axioms underlying the theory of probability. This is done by application of the updating rule of Bayes theorem, which becomes very controversial when one considers the estimation of statistical parameters from the point of view of the classical, frequentist statistics or of the Bayesian, subjectivist statistics (Chapter 9).

Example 4.4 [1]

Consider a pile foundation, in which pile groups are used to support the individual column footings. Each of the pile group is designed to support a load of 200 tons. Under normal condition, this is quite safe. However, on rare occasions the load may reach as high as 300 tons. The foundation engineer wishes to know the probability that a pile group can carry this extreme load of up to 300 tons.

Based on previous experience with similar pile foundations, supplemented with blow counts and soil tests, the engineer estimated a probability of 0.70 that any pile group can support a 300-ton load. Also, among those that have capacity less than 300 tons, 50% failed at loads less than 280 tons.

To improve the estimated probability, the foundation engineer orders one pile group to be proof-loaded to 280 tons.

1. If the pile group survives the specified proof load, what is the probability that the pile group can support a load of 300 tons?

Solution

Let

$$A = \text{event that the capacity of pile group} \geq 300 \text{ tons}$$
$$T = \text{event of a successful proof load.}$$

Then, according to the information given above, $P(\overline{T} \mid \overline{A}) = 0.5$, $P(A) = 0.70$ and $P(T \mid A) = 1$. Bayes' theorem then gives:

$$P(A \mid T) = \frac{P(T \mid A)P(A)}{P(T \mid A)P(A) + P(T \mid \overline{A})P(\overline{A})} = \frac{(1.00)(0.70)}{(1.00)(0.70) + (0.5)(0.3)} = 0.824$$

Therefore, if the proof test is successful, the required probability is increased from 0.70 to 0.824.

4.5   Random variables

The outcome ω of a random experiment in the sample space Ω can be described by a real *random variable* $X(\omega) \in \mathfrak{R}$. For example, we can describe any event associated with the outcomes of an experiment of rolling a dice by a real variable X in $\mathfrak{R}$. For a given numerical value x we can then define the event described by all possible outcomes associated to values of the random variable X less than x: for example, for $x = 4.72$ the event $\{X \leq 4.72\}$ corresponds to the union of the outcomes $\{1 \cup 2 \cup 3 \cup 4\}$; the event $\{X \leq 0\}$ is the null set since the outcomes of the roll of the dice are not associated to any negative value of X; for $x = \infty$ the event $\{X \leq \infty\}$ is the full sample space Ω.

By establishing a univocal mapping between the outcomes of a random experiment and the values of a random variable, one can handle the uncertain events in terms of their mathematical abstractions, sparing the need of an actual word description for each particular physical phenomenon. In other words, general mathematical models of random behaviours can be built which apply to different physical phenomena which behave similarly.

4.5.1 Probability functions

Cumulative distribution function

The *cumulative distribution function* (cdf) $F_X(x)$ of the random variable X gives the probability of the event $\{X \le x\}$ for any numerical value x. From the definition, the following properties of $F_X(x)$ hold:

- $\lim\limits_{x \to -\infty} F_X(x) = 0$
- $\lim\limits_{x \to +\infty} F_X(x) = 1$
- $F_X(x)$ is a non-decreasing function of x
- The probability that X takes on a value in the interval $[a,b]$ is
  $$P\{a < X \le b\} = F_X(b) - F_X(a)$$

Probability mass function (discrete random variables)

Consider a random variable X which can take on only discrete values $x_i, i = 1,2,...n$. The discrete function of the probability values p_i with which X takes on the values $x_i, i = 1,2,...n$, is termed *probability mass function* (pmf) and gives a more detailed information on the behaviour of the random variable.

The corresponding cumulative distribution function is given by

$$F_X(x) = \sum_{x_i \le x} p_i \qquad (4.19)$$

Probability density function (continuous random variables)

Consider a random variable X which can take on continuous values x in $\Re$, with cdf $F_X(x)$. As mentioned in Section 4.3.4, continuous intervals cannot be constructed by adding elementary points in a countable manner and correspondingly, probabilities of continuous intervals cannot be assigned by the addition law of probability. Thus, the probability of X taking on a particular value x is zero. Instead, we can consider a small interval dx centred around the value x and consider the probability of the random variable X taking any value within such interval:

$$P\{x \le X < x + dx\} = F_X(x + dx) - F_X(x) = f_X(x)dx \qquad (4.20)$$

where $f_X(x)$ is the so-called *probability density function* (pdf) of X.
Taking the limit for the interval dx becoming infinitesimal,

$$f_X(x) = \lim_{dx \to 0} \frac{F_X(x + dx) - F_X(x)}{dx} = \frac{dF_X}{dx} \qquad (4.21)$$

Note that $f_X(x)$ is not a probability but a probability per unit of x, i.e. a probability density: when multiplied by dx it becomes the probability of X falling in the interval $[x, x + dx)$.

4.5.2   Summary measures: percentiles, median, mean, variance

The cumulative distribution and probability mass and density functions give the most complete description of the behaviour of a random variable. Yet, in engineering practice a pointwise description of the location and shape of such probability distributions is often needed. For this reason, various *summary measures* can be adopted.

Distribution Percentiles

The α *percentile* of the distribution $F_X(x)$ is the value x_α at which

$$F_X(x_\alpha) = \frac{\alpha}{100} \qquad (4.22)$$

In particular, the *50-th* percentile x_{50} is called the *median* of the distribution and it represents the numerical value for which there is a symmetric probability of 0.5 that the random variable X takes values below or above, i.e.

$$F_X(x_{50}) = 0.5 \qquad (4.23)$$

In other words, half of the probability mass lies below x_{50} and half above.

Mean

The *mean* of the distribution $F_X(x)$ provides information as to where the probability distribution is located on $\mathfrak{R}$, i.e. where the probability mass is concentrated on average. It is often referred to as *expected value* of the distribution, defined as

$$\mu_X = E[X] = \;<X> \; = \sum_{i=1}^{n} x_i p_i \qquad (discrete\ random\ variables)$$
$$\qquad\qquad\qquad (4.24)$$
$$= \int_{-\infty}^{\infty} x f_X(x) dx \qquad (continuos\ random\ variables)$$

Central moments

The *central moments* of the distribution $F_X(x)$ provide information on its shape relative to the mean. In general, the *n*-th central moment of the distribution is defined as:

$$\sigma_X^n = \sum_i (x_i - \mu_X)^n p_i \qquad (discrete\ random\ variables)$$
$$\qquad\qquad\qquad (4.25)$$
$$= \int_{-\infty}^{\infty} (x - \mu_X)^n f_X(x) dx \qquad (continuos\ random\ variables)$$

Often used are the second and third moments, $n=2$ and 3 respectively. The former (σ_X^2), called *variance* and often indicated also as $Var[X]$, gives a measure of the spread of the distribution around the mean: the larger it is, the more the distribution is spread out over $\Re$ around the mean; the smaller it is, the more the distribution is peaked on the mean value. The latter (σ_X^3) is called *kurtosis* and gives a measure of asymmetry: a value close to zero indicates a fairly symmetric distribution; negative values indicate that the distribution is skewed to the right (i.e. values smaller than the mean are more dispersed in a large tail); positive values indicate that the distribution is skewed to the left (i.e. values larger than the mean are more dispersed in a large tail).

Finally, a combined measure of spread and location, called *coefficient of variation (Cov)* is often used in civil engineering:

$$Cov_X = \frac{\sigma_X}{\mu_X} \tag{4.26}$$

where σ_X is the square root of the variance and is called *standard deviation* (often also indicated as $Std[X]$).

Chebychev's inequality

Chebychev's inequality provides an estimate of the probability of dispersion around the mean of the values of a random variable X with distribution $F_X(x)$. From the definition of the variance, we have:

$$\sigma_X^2 = \int_{-\infty}^{\infty}(x-\mu_X)^2 f_X(x)dx \geq \int_{|x-\mu_X|\geq k\sigma_X}(x-\mu_X)^2 f_X(x)dx \geq k^2\sigma_X^2 \int_{|x-\mu_X|\geq k\sigma_X} f_X(x)dx =$$

$$= k^2\sigma_X^2 P\left(|x-\mu_X| \geq k\sigma_X\right)$$

$$\sigma_X^2 = \int_{-\infty}^{\infty}(x-\mu_X)^2 f_X(x)dx \geq \int_{|x-\mu_X|\geq k\sigma_X}(x-\mu_X)^2 f_X(x)dx \geq k^2\sigma_X^2 \int_{|x-\mu_X|\geq k\sigma_X} f_X(x)dx =$$

$$= k^2\sigma_X^2 P\left(|x-\mu_X| \geq k\sigma_X\right)$$

(4.27)

from which it follows that

$$P(|x-\mu_X| \geq k\sigma_X) \leq \frac{1}{k^2}$$

(4.28)

Example 4.5 (discrete random variable) [1]

A contractor is planning the purchase of equipment, including bulldozers, needed for a new project in a remote area. Suppose that from his previous experience, he figures there is a 50% chance that each bulldozer can last at least 6 months without any breakdown.

1.   If he purchased 3 bulldozers, what is the probability that there will be only 1 bulldozer left operative in 6 months?
2.   Let X be the random variable whose values represent the number of good bulldozers after 6 months. The probability that a bulldozer will remain operational after 6 months is $p = 0.8$. Using the above information, plot the probability mass function (pmf) as well as the cumulative distribution function (cdf) of X.
3.   Using information from part 2., compute the following:
     (i)   Mean of X
     (ii) Variance of X
     (iii) Standard Deviation of X
     (iv) Coefficient of Variation of X

Solution

4.5.1   Let

G = event where a Bulldozer is in good condition.
B = event where a Bulldozer is in bad condition.

The possible statuses of the three bulldozers would be:

{GGG, GGB, GBB, BBB, BGG, BBG, GBG, BGB}

In this case, there are a total of 8 possibilities. Since the condition of a bulldozer is equally likely to be good or bad, the 8 possible statuses of the 3 bulldozers are also equally likely to occur. The events of interest are *GBB, BBG, BGB*. Therefore, the probability of having only 1 bulldozer left operative in 6 months is simply = *3/8 = 0.375*

4.5.2　　The possible values of *X* are *{0, 1, 2, 3}*. Then,

$$P_X (0) = (1 - p)^3 = (0.2)^3 = 0.008$$
$$P_X (1) = 3p(1 - p)^2 = 3(0.8)(0.2)^2 = 0.096$$
$$P_X (2) = 3p^2(1 - p) = 3(0.8)^2(0.2) = 0.384$$
$$P_X (3) = p^3 = (0.8)^3 = 0.512$$

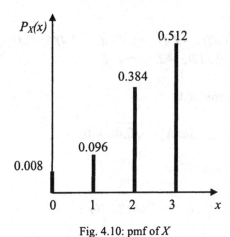

Fig. 4.10: pmf of *X*

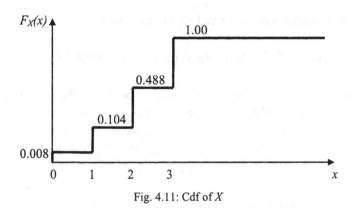

Fig. 4.11: Cdf of X

4.5.3

(i) Mean of X

$$E[X] = 0(0.008) + 1(0.096) + 2(0.384) + 3(0.512) = 2.40$$

(ii) Variance of X

$$Var[X] = 0.008(0-2.4)^2 + 0.096(1-2.4)^2 + 0.384(2-2.4)^2 + \\ + 0.512(3-2.4)^2 = 0.48$$

(iii) Standard Deviation of X

$$Std[X] = \sqrt{0.48} = 0.69$$

(iv) Coefficient of Variation of X

$$Cov_X = (0.69)/(2.40) = 0.29$$

Example 4.6 (continuous random variable) [1]

Suppose that a random variable X has a pdf of the form (Fig. 4.12):

$$f_X(x) \quad = \alpha x^2 \qquad\qquad 0 \le x \le 10$$
$$= 0 \qquad\qquad \text{elsewhere}$$

1. Under what condition (i.e. what value of α) is this function a bona fide pdf?
2. What is $P(X > 5)$?
3. Compute the following:
    (i)   Mean of X
    (ii) Variance of X
    (iii) Standard Deviation of X
    (iv) Coefficient of Variation of X
    (v)   Median of X

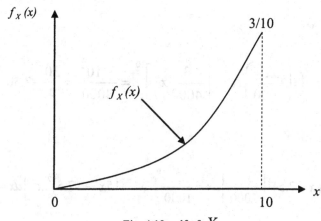

Fig. 4.12: pdf of X

Solution

4.6.1

In order to satisfy all the properties of a pdf, we must have

$$\int_0^{10} \alpha x^2 dx = 1$$

from which we get that

$$\frac{\alpha}{3}(10)^3 = 1.0$$

Therefore, solving for α, we get

$$\alpha = (3/1000)$$

4.6.2

$$P(X > 5) = 1 - P(X \leq 5) = 1 - \int_0^5 \frac{3x^2}{1000} dx = 1 - \frac{5^3}{1000} = 0.875$$

4.6.3

(i) Mean of X

$$E[X] = \int_0^{10} x \left(\frac{3x^2}{1000} \right) dx = \left[\frac{3}{4000} x^4 \right]_0^{10} = \frac{3 \cdot 10^4}{4000} = \frac{30}{4} = 7.50$$

(ii) Variance of X

$$Var[X] = \int_0^{10} (x - 7.5)^2 \left(\frac{3x^2}{1000} \right) dx = \frac{3}{1000} \int_0^{10} \left[x^4 - 15x^3 + (7.5)^2 x^2 \right] dx = 3.75$$

(iii) Standard Deviation of X

$$Std[X] = \sqrt{3.75} = 1.94$$

(iv) Coefficient of Variation of X

$$Cov_X = (1.94)/(7.50) = 0.26$$

(v) Median of X

From Fig. 4.12, the modal value is obviously $\tilde{x} = 10$. To determine the median, one must solve

$$\int_0^{x_m} \frac{3x^2}{1000} dx = 0.50$$

from which we get

$$x_m^3 = 500$$

and thus the median is

$$x_m = 7.94.$$

4.5.3 The hazard function

Continuous random variables are often used in risk and reliability analyses. Of particular importance is the time to failure of a component T whose cdf $F_T(t)$ and pdf $f_T(t)$ are typically called the failure probability and density functions at time t. The complementary cumulative function (ccdf) $R(t) = 1 - F_T(t) = P(T > t)$ is called *reliability* or *survival function* of the component at time t and gives the probability that the component survives up to time t with no failures.

Another information of interest for monitoring the failure evolution process of a component is given by the probability that it fails in an interval dt knowing that it has survived with no failures up to the time of beginning of the interval, t. This probability is expressed in terms of the product of the interval dt times a conditional probability density called *hazard function* or *failure rate* and usually indicated by the symbol $h_T(t)$:

$$h_T(t)dt = P(t < T \le t + dt \mid T > t) = \frac{P(t < T \le t + dt)}{P(T > t)} = \frac{f_T(t)dt}{R(t)} \quad (4.29)$$

The hazard function $h_T(t)$ gives the same information of the pdf and cdf to whom it is univocally related by eq. (4.29) and its integration, i.e.

$$F_T(t) = 1 - e^{-\int_0^t h_T(s)ds} \qquad (4.30)$$

Fig. 4.13 shows the most common patterns of evolution of $h_T(t)$ encountered in practice [2].

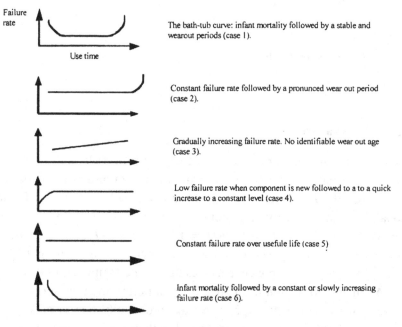

Fig. 4.13: Patterns of time evolution of the hazard function (or failure rate) [2]

In principle, the hazard function follows the so called 'bath-tub' curve (Fig. 4.13, case 1) which shows three distinct phases in the life of a component: the first phase corresponds to a failure rate decreasing with time and it is characteristic of the *infant mortality* or *burn in* period whereupon the more the component survives, the lower becomes its probability of failure (this period is central for warranty analysis); the second period, called *useful life*, corresponds to a failure rate independent of time: during this period, failures occur at random times with no influence on the usage time of the component; finally, the last period sees an increase in the failure rate with time and corresponds to the development of irreversible aging processes which make the component more and more prone to fail as time goes by.

Deviations from this general behaviour (cases 2-6 in Figure 4.13) may occur, depending on the burn-in and maintenance procedures adopted by the particular industry.

4.6 Probability distributions

A number of classes of stochastic processes can be described mathematically in terms of special analytical forms of the pdf and cdf.

4.6.1 Univariate discrete distributions

Binomial Distribution

Consider n independent realizations (trials) of the stochastic experiment known as *Bernoulli process*, described by a discrete random variable Y with only two possible outcomes: 1 (success), with probability p and 0 (failure), with probability $1 - p$.

Let X be the discrete random variable describing the number of successes (realizations of the outcome 1) out of the n trials, independently of the sequence with which the successes appear. The sample space of X comprises all discrete values from 0 to n.

The random variable X is related to the random variables Y_i, $i = 1,2,...n$, describing the individual Bernoulli trials as follows:

$$X = \sum_{i=1}^{n} Y_i \tag{4.31}$$

The distribution of the discrete random variable X above defined is called *Binomial*. Its probability mass function $b(k;n,p)$ gives the probability of obtaining k successes out of n Bernoulli trials when the probability of success in the individual trial is p :

$$b(k; n, p) = \binom{n}{k} p^k (1-p)^{n-k} \qquad k = 1, 2, \ldots n \qquad (4.32)$$

The expected value and variance of the distribution are:

$$
\begin{aligned}
E[X] &= np \\
Var[X] &= np(1-p)
\end{aligned}
\qquad (4.33)
$$

Geometric Distribution

Considering the previous problem setting of independent trials of the stochastic experiment known as *Bernoulli process*, we focus now on the probability that the first success occurs at the t - th trial.

Only one specific sequence is now considered, i.e. that with all failures in the first $t-1$ trials (each one occurring with probability $1-p$) and a success at the t-th trial (which occurs with probability p).

The distribution of the corresponding random variable is called *Geometric*. Its probability mass function is

$$g(t; p) = (1-p)^{t-1} p \qquad t = 1, 2, \ldots \qquad (4.34)$$

Note that (4.34) is also the distribution of the number of trials between two successive occurrences of success (realizations of 1), since the Bernoulli trials are independent and the probability of success p remains the same in all trials.

The expected value of the geometric distribution is computed as follows:

$$E[T] = \sum_{t=1}^{\infty} t(1-p)^{t-1} p = p[1 + 2(1-p) + 3(1-p)^2 + \ldots] = \frac{p}{[1-(1-p)]^2} = \frac{1}{p}$$

$$(4.35)$$

This quantity is often called the *return period* of the stochastic process.

Poisson Distribution

Consider now stochastic events that occur in a continuum period (e.g. the number of earthquakes which occur in a given region over a given period of time, the number of cars crossing a given intersection over a given period of time, the number of failures of a given type of component over a given period of time).

The rate of occurrence λ of the events is assumed constant and the events are assumed independent of each other.

The distribution of the discrete random variable K describing this process is called *Poisson*. Its probability mass function gives the probability that k events occur in the period of observation $(0, t)$ and is defined as:

$$p(k; (0, t), \lambda) = \frac{(\lambda t)^k}{k!} e^{-\lambda t} \qquad k=1, 2, \ldots \qquad (4.36)$$

The expected value and variance of the distribution are:

$$E[K] = \lambda t$$
$$Var[K] = \lambda t \qquad (4.37)$$

As it can be intuitively understood, the Poisson distribution derives from the binomial one in the limit for $p \to 0, n \to \infty$ so that the product $np = \lambda t$ remains constant.

Example 4.7 (Poisson distribution) [1]

On the average two damaging earthquakes occur in a certain country every 5 years. Assume the occurrence of earthquakes is a Poisson process in time. For this country, compute the following:

1. Determine the probability of getting 1 damaging earthquake in 3 years.
2. Determine the probability of no earthquakes in 3 years.
3. What is the probability of having at most 2 earthquakes in one year?
4. What is the probability of having at least 1 earthquake in 5 years?

Solution

$$\lambda = \frac{2}{5} = 0.4 y^{-1}$$

4.7.1 $P(1$ earthquake in 3 years$) = p(1;(0,3),0.4) = (0.4t)e^{-0.4t}\big|_{t=3} = 0.3614$

4.7.2 $P(0$ earthquake in 3 years$) = p(0;(0,3),0.4) = \frac{(0.4t)^0}{0!}e^{-0.4t}\big|_{t=3} = 0.3$

4.7.3 $P(K \leq 2$ in 1 year$) = \sum_{k=0}^{2} p(x;(0,3),0.4) = p(0;(0,3),0.4) + p(1;(0,3),0.4)$

$$= e^{-0.4t} + 0.4t(e^{-0.4t}) + \frac{(0.4t)^2}{2}e^{-0.4t}\big|_{t=1} = 0.992$$

4.7.4 $P(K \geq 1$ in 5 years$) = 1 - P(K = 0$ in 5 years$) = 1 - e^{-0.4t}\big|_{t=5} = 1 - e^{-2} = 0.864$

Example 4.8 (Binomial and Poisson distribution) [1]

The occurrences of floods may be modelled by a Poisson process with rate υ. Let p(k; t, υ) denote the probability of k flood occurrences in t years.

1.   If the mean occurrence rate of floods for a certain region A is once every 8 years, determine the probability of no floods in a 10-year period; of 1 flood; of more than 3 floods.
2.   A structure is located in region A. The probability that it will be inundated, when a flood occurs, is 0.05. Compute the probability that the structure will survive if there are no floods; if there is 1 flood; if there are n floods. Assume statistical independence between floods.
3.   Determine the probability that the structure will survive over the 10-year period.

Solution

4.8.1 $\upsilon = \dfrac{1}{8} = 0.125 \text{ y}^{-1}$

$P\{K = 0 \text{ in } 10 \text{ years}\} = e^{-1.25} = 0.286$

$P\{K = 1 \text{ in } 10 \text{ years}\} = 1.25(e^{-1.25}) = 0.3525$

$P\{K > 3 \text{ in } 10 \text{ years}\} = 1 - P\{K \le 3 \text{ in } 10 \text{ years}\} = 1 - \displaystyle\sum_{k=0}^{3} p(k; t = 10, \upsilon = 0.125)$

$$= 1 - 0.286 - 0.3575 - \dfrac{1.25^2}{2!} e^{-1.25} - \dfrac{1.25^3}{3!} e^{-1.25} = 0.0394$$

4.8.2 $P\{\text{structure fails} \mid \text{flood}\} = 0.05$

$P\{\text{structure survives} \mid \text{flood}\} = 0.95$

$P\{\text{structure survives, 0 flood}\} = P\{0 \text{ flood}\} \, P\{\text{structure survives} \mid 0 \text{ flood}\}$
$= 0.286 \, (1) = 0.286$

$P\{\text{structure survives, 1 flood}\} = P\{1 \text{ flood}\} \, P\{\text{structure survives} \mid 1 \text{ flood}\}$
$= (0.3525)(0.95)$
$= 0.3396$

$P\{\text{structure survives, } n \text{ independent floods}\} =$
$= P\{n \text{ floods}\} \, P\{\text{structure survives} \mid n \text{ independent floods}\}$
$$= \left[\dfrac{1.25^n}{n!} \right] \cdot 0.95^n$$

4.8.3 $P\{\text{structure survives over } 10 \text{ years}\} = \displaystyle\sum_{n=0}^{\infty} \dfrac{1.25^n}{n!} e^{-1.25} (0.95)^n = \sum_{n=0}^{\infty} \dfrac{1.1875^n}{n!} e^{-1.25}$

$$= e^{1.1875 - 1.25} = e^{-0.0625}$$
$$= 0.9394$$

4.6.2   Univariate continuous distributions

Exponential Distribution

Consider a component operating at time $t = 0$ and characterized by a constant failure rate $h_T(t) = \lambda$. Let us consider a given time t and subdivide the time period $(0, t)$ in n subintervals of equal length Δt. In each subinterval, the component either survives with constant survival probability equal to $1 - \lambda \Delta t$ or fails with complementary probability $\lambda \Delta t$. Hence, each interval represents a Bernoulli trial. Then, the probability that the component has no failures up to time t is given by the binomial distribution (4.32) for the discrete random variable *number of failures*, evaluated at $k = 0$,

$$b(0; n, \lambda\Delta t) = \binom{n}{0}(\lambda\Delta t)^0(1 - \lambda\Delta t)^{n-0} = (1 - \lambda\Delta t)^n \qquad (4.38)$$

In (4.38) the random variable is the number of failures in n Bernoulli trials. Looking at the stochastic process in terms of the continuous random variable *failure time T*, the reliability of the component at time t, i.e. the probability that the component does not fail up to time t, is the probability that T takes on values larger than,

$$R(t) = P(T > t) = \lim_{\substack{n\to\infty \\ \Delta t\to\infty}} (1 - \lambda\Delta t)^n = \lim_{n\to\infty}\left(1 - \lambda\frac{t}{n}\right)^n = e^{-\lambda t} \qquad (4.39)$$

The cdf of T is then:

$$F_T(t) = P(T \le t) = 1 - e^{-\lambda t} \qquad (4.40)$$

with corresponding pdf (Fig. 4.14):

$$\begin{aligned} f_T(t) &= \lambda e^{-\lambda t} && t \ge 0 \\ &= 0 && t < 0 \end{aligned} \qquad (4.41)$$

and hazard function:

$$\begin{aligned} h_T(t) &= \frac{f_T(t)}{R(t)} = \lambda && t \ge 0 \\ &= 0 && t < 0 \end{aligned} \qquad (4.42)$$

Such distribution is called *exponential* and it is the only distribution characterized by a constant hazard rate. For this reason it is widely used in reliability practice to describe the flat, constant part (useful life) of the bath-tub hazard function of a component (section 4.5.3).

The expected value and variance of the distribution are:

$$E[T] = \frac{1}{\lambda}$$
$$Var[T] = \frac{1}{\lambda^2}$$

(4.43)

When T is the time to failure, the expected value represents the return period of failures and is often called *Mean-Time-To-Failure* (MTTF).

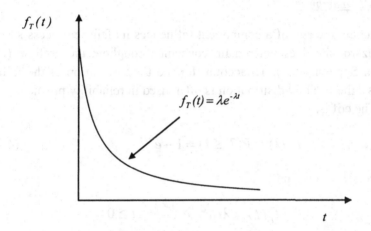

Fig. 4.14: Exponential distribution

Note that the probability of the failure time T being larger than a given value t is equal to the probability of having 0 failures in the period $(0, t)$: in the case of constant failure rate λ, the former is given by the exponential complementary cumulative distribution function (4.39), whereas the latter is given by the Poisson distribution (4.36) for $k = 0$.

Finally, when the failure rate is constant, the process is said to be *memoryless*. Indeed, suppose that a component with constant failure rate λ is found still operational at a given time t_1 and that one is interested in the probability of its failure before $t_2 > t_1$. This is given by the following conditional probability:

$$P(t_1 < T < t_2 \mid T > t_1) = \frac{P(t_1 < T < t_2)}{P(T > t_1)} = \frac{F(t_2) - F(t_1)}{R(t_1)} = \frac{e^{-\lambda t_1} - e^{-\lambda t_2}}{e^{-\lambda t_1}} = 1 - e^{-\lambda(t_2 - t_1)}$$

$$(4.44)$$

The distribution of the failure times starting from t_1 is still exponential with failure rate λ, so that knowing that the component has survived with no failures up to t_1 does not change the probability of its failure within the next interval of duration $(t_2 - t_1)$.

Weibull Distribution

In practice, the age of a component influences its failure process so that the hazard rate does not remain constant throughout the lifetime (Fig. 4.13 in Section 4.5.3). To account for the time evolution of the failure process, the Weibull distribution is often used in reliability practice.

The cdf is:

$$F_T(t) = P(T \le t) = 1 - e^{-\lambda t^\alpha} \tag{4.45}$$

with corresponding pdf:

$$
\begin{aligned}
f_T(t) &= \lambda \alpha t^{\alpha-1} e^{-\lambda t^\alpha} && t \ge 0 \\
&= \lambda \alpha t^{\alpha-1} e^{-\lambda t^\alpha} && t < 0
\end{aligned}
\tag{4.46}
$$

The expected value and variance of the Weibull distribution are:

$$E[T] = \frac{1}{\lambda} \Gamma\!\left(\frac{1}{\alpha} + 1\right)$$

$$Var[T] = \frac{1}{\lambda^2}\left(\Gamma\!\left(\frac{2}{\alpha} + 1\right) - \Gamma\!\left(\frac{1}{\alpha} + 1\right)\right)^2 \tag{4.47}$$

where the Gamma function $\Gamma(\cdot)$ is the generalization to non-integer numbers of the factorial and is defined as

$$\Gamma(k) = \int_0^\infty x^{k-1}e^{-x}dx \qquad k > 0 \tag{4.48}$$

which by integration by parts yields,

$$\Gamma(k) = (k-1)\Gamma(k-1) \tag{4.49}$$

Normal or Gaussian Distribution

The importance of the *normal* or *Gaussian* distribution is related to the famous *central limit theorem*: for any distribution of independent random variables X_i, their sum $X_1 + X_2 + ... + X_n$ is a random variable which for large n tends to be distributed as a normal distribution. This, for example, justifies the use of the normal distribution to describe experimental errors which are typically the effect of several independent random phenomena.

The Gaussian distribution is the only distribution with a symmetric, bell shape. Its pdf is

$$f_X(x; \mu_X, \sigma_X) = \frac{1}{\sqrt{2\pi}\sigma_X} e^{-\frac{1}{2}\left(\frac{x-\mu_X}{\sigma_X}\right)^2} \qquad -\infty < x, \mu_X < \infty; \sigma_X > 0 \tag{4.50}$$

The expected value and variance of the Gaussian distribution are:

$$E[X] = \mu_X \tag{4.51}$$
$$Var[X] = \sigma_X^2$$

A random variable distributed as a normal with mean μ_X and standard deviation σ_X is typically indicated as $X \sim N(\mu_X, \sigma_X)$. Often in practice, one refers to the so called *standard normal variable* $\xi = \dfrac{X - \mu_X}{\sigma_X} \sim N(0,1)$ which is easily tabulated (Appendix A).

Log-normal Distribution

Let us consider a stochastic process of a random variable X which, beginning from an initial value $x_0 > 0$ is increased by successive random, independent contributions proportional to the current value of X. Let $\{r\}$ be the sequence of independent random variables such that:

$$x_{i+1} = x_i + r_i x_i \qquad (4.52)$$

Assuming small relative increases $\Delta x_i = \dfrac{x_{i+1} - x_i}{x_i}$:

$$r_1 + r_2 + ... + r_n = \frac{\Delta x_1}{x_1} + \frac{\Delta x_2}{x_2} + ... + \frac{\Delta x_n}{x_n} \qquad (4.53)$$

which for large number n becomes:

$$\lim_{n \to \infty}(r_1 + r_2 + ... + r_n) = \lim_{n \to \infty}\left(\frac{\Delta x_1}{x_1} + \frac{\Delta x_2}{x_2} + ... + \frac{\Delta x_n}{x_n} \right) = \int_{x_0}^{x_n} \frac{du}{u} = \ln \frac{x_n}{x_0}$$

$$(4.54)$$

For the central limit theorem, the first limit tends to a normal random variable and so does $Z = \ln X$, with $X = \lim_{n \to \infty} \frac{x_n}{x_0}$. The distribution of X is called *log-normal*. In other words, X is log-normal if $Z = \ln X$ is a normal random variable.

Denoting by $f_Z(\cdot)$ and $g_X(\cdot)$ the pdfs of the normal random variable Z (4.52) and of the log-normal random variable X, respectively, from

$$f_Z dz = g_X dx \qquad (4.55)$$

one obtains the log-normal pdf:

$$g_X(x;\mu_Z,\sigma_Z) = \frac{1}{\sqrt{2\pi}\sigma_Z}\frac{1}{x}e^{-\frac{1}{2}\left(\frac{\ln x-\mu_Z}{\sigma_Z}\right)^2} \qquad x,\sigma_Z > 0 \quad (4.56)$$

where μ_Z and σ_Z are the mean and standard deviation of the corresponding normal distribution of $Z = \ln X$. The expected value and variance of $g_X(x;\mu_Z,\sigma_Z)$ are:

$$E[X] = e^{\mu_Z+\frac{\sigma_Z^2}{2}}$$
$$Var[X] = e^{2\mu_Z+\sigma_Z^2}\left(e^{\sigma_Z^2} - 1\right) \qquad (4.57)$$

The log-normal distribution is asymmetric, skewed to the right and it is often used to represent the uncertainty in the estimates of the components failure rates. In this view, it is often characterized in terms of percentiles and the error factor

$$\frac{x_{95}}{x_{50}} = \frac{x_{50}}{x_5} = e^{\xi_{95}\sigma_Z} \qquad (4.58)$$

where $\xi \sim N(0,1)$.

Example 4.9 (exponential and Gaussian distribution) [1]

The daily concentration of a certain pollutant in a stream has the exponential distribution shown in Fig. 4.15.

1. If the mean daily concentration of the pollutant is 2 mg/10^3 liter, determine the constant c in the exponential distribution.
2. Suppose that the problem of pollution will occur if the concentration of the pollutant exceeds 6mg/10^3 liter. What is the probability of a pollution problem resulting from this pollutant in a single day?
3. What is the return period (in days) associated with this concentration level of 6 mg/10^3 liter? Assume that the concentration of the pollutant is statistically independent between days.
4. What is the probability that this pollutant will cause a pollution problem at most once in the next 3 days?

5.   If instead of the exponential distribution, the daily pollutant concentration is Gaussian with the same mean and variance, what would be the probability of pollution in a day in this case?

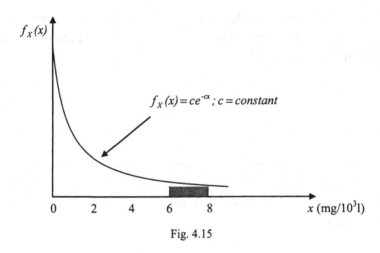

Fig. 4.15

Solution

4.9.1   First, we verify the normalization of the probability density function, i.e.

$$\int_0^\infty ce^{-cx}dx = 1 \Rightarrow -e^{-cx}\Big|_0^\infty = 1$$

Then, from the expected value of the exponential distribution we have:

$$E[X] = 1/c = 2 \Rightarrow c = 0.5$$

$$E[X^2] = \int_0^\infty (x^2)0.5e^{-0.5x}dx = 8 \qquad \sigma_X^2 = E[X^2] - E^2[X] = 8 - 4 = 4$$

4.9.2 $P(\text{pollution}) = P(X > 6) = 1 - P(X \le 6) = 1 - \int_0^6 0.5e^{-0.5x}dx = 1 + e^{-0.5x}\Big|_0^6 = 0.0498$

For simplicity of notation, we shall denote $P(X > 6)$ by $p_{X>6} = 0.0498$.

4.9.3 $E[T_{X>6}] = \dfrac{1}{p_{X>6}} = \dfrac{1}{0.0498} = 20$ days

4.9.4 P(pollution at most once in 3 days) $= \sum_{k=0}^{1} \binom{3}{k} p_{X>6}^{k} (1 - p_{X>6})^{3-k}$

$$= (1 - 0.0498)^3 + (3)(0.0498) (1 - 0.0498)^2$$

$$= 0.993$$

4.9.5 $P(X > 6) = 1 - P(X \leq 6) = 1 - P(\xi \leq 2) = 1 - \Phi(2) = 1 - 0.977 = 0.023$
where $\bar{\xi} \sim N(0,1)$ is the standard normal variable (Appendix A).

Example 4.10 (Gaussian and Log-normal distributions) [1]

A contractor estimates that the expected time for completion of job A is 30 days. Because of uncertainties that exist in the labor market, materials supply, bad weather conditions, and so on, he is not sure that he will finish the job in exactly 30 days. However, he is 90% confident that the job will be completed within 40 days. Let X denote the number of days required to complete job A.

1. Assume X to be a Gaussian random variable; determine μ and σ and also the probability that X will be less than 50, based on the given information.
2. Recall that a Gaussian random variable ranges from $-\infty$ to $+\infty$. Thus X may take on negative values that are physically impossible. Determine the probability of such an occurrence. Based on this result, is the assumption of the normal distribution for X reasonable?
3. Let us now assume that X has a log-normal distribution with the same expected value and variance as those in the normal distribution of part (1). Determine the parameters μ_z and σ_z, and also the probability that X will be less than 50. Compare this with the result of part (1).

Solution

4.10.1 Let X be the number of days required to complete job A.

$$E[X] = \mu_X = 30 \qquad X \sim N(30, \sigma_x)$$

$$P(X \leq 40) = 0.9 \Rightarrow P(\frac{x-30}{\sigma_X} \leq \frac{40-30}{\sigma_X}) = 0.9 \Rightarrow P(\xi \leq \frac{10}{\sigma_X}) = 0.9$$

From the tabulated values $\Phi(\xi)$ of the normal standard variable ξ (Appendix A):

$$\Phi(\frac{10}{\sigma_X}) = 0.9 \Rightarrow \frac{10}{\sigma_X} = \Phi^{-1}(0.9) = 1.29 \Rightarrow \sigma_X = 7.752$$

Therefore $P(X \leq 50) = P(\xi \leq 2.58) = \Phi(2.58) = 0.995$

4.10.2 The limit value is $X = 0$ which corresponds to the following value of the standard normal variable:

$$\bar{\xi} = \frac{0-30}{7.752} = -3.87$$

Then, the probability of negative values is:

$$P(\xi \leq \bar{\xi}) = 1 - \Phi(3.87) = 1 - 0.999946 = 5.4 \cdot (10^{-5})$$

which is negligible: the assumption of normal distribution is acceptable.

4.10.3 $X \sim f_X(x; \mu_z, \sigma_z)$ \qquad with $\mu_x = 30$ and $\sigma_x = 7.752$

$$\mu_X = e^{\mu_z + \frac{1}{2}\sigma_z^2}$$

$$\sigma_z^2 = \ln\left(1 + \frac{\sigma_x^2}{\mu_x^2}\right) = 0.0646 \qquad\qquad \mu_z = \ln \mu_x - \frac{1}{2}\sigma_z^2 = 3.3688$$

Therefore, $X \sim f_X(x; 3.3688, 0.0646)$

$$P(X \leq 50) = P\left(\frac{\ln X - \mu_z}{\sigma_z} \leq \frac{\ln 50 - \mu_z}{\sigma_z}\right) = P(\xi \leq 2.137) = \Phi(2.137) = 0.983$$

4.7 Regression and correlation analyses

When dealing with two or more variables, the functional relation between the variables is often of interest. However, if one or more variables are random, for a given value of one variable (the controlled variable), there is a range of possible values of the others and thus a probabilistic description is required.

If the probabilistic relationship between the variables is described in terms of the mean and variance of one random variable as a function of the other variables, we have what is known as *"regression analysis"*. When the analysis is limited to linear mean value functions, it is called *"linear regression"*. In general, however, regression may be nonlinear.

4.7.1 Regression with constant variance

Considering pairwise data of two variables, X and Y, the possible value of one variable, e.g. Y, may depend on the values of the other variable X. For this reason, it would be inappropriate to analyse the data for Y (e.g. in determining the mean and variance of Y) without due consideration of X. In the case of Fig. 4.16, we observe that there is a

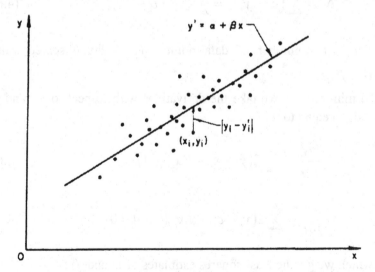

Fig. 4.16: Linear regression analysis of data for two variables

general tendency for the values of Y to increase with increasing values of X (X may be deterministic or random). Hence, the mean value of Y will also increase with increasing values of X although, due to the stochastic behavior of the process, the actual values of Y may not always increase with increasing values of X.

In general, the mean value of Y will depend on the value of X. Suppose that this relationship is linear; that is,

$$E[Y|X = x] = \alpha + \beta x \qquad (4.59)$$

where α, β are constants and the variance of Y may be independent or a function of X. This is known as the linear regression of Y on X.

Let us consider the case with $Var[Y|X = x] = constant$.

Depending on the values of α and β, there are many straight lines that could represent the function $E[Y|X = x]$ in the light of the available data. The 'best' line is that which passes through the data points with the least error. The coefficient of this line with least total error can be found by minimizing the sum of the squared errors:

$$\Delta^2 = \sum_{i=1}^{n} (y_i - y_i')^2 = \sum_{i=1}^{n} (y_i - \alpha - \beta x_i)^2 \qquad (4.60)$$

where n is the number of data points, y_i is the observed value, $y_i' = \alpha + \beta x_i$.

To minimize Δ^2 we take the derivatives with respect to α and β and set them equal to 0:

$$\frac{\partial \Delta^2}{\partial \alpha} = \sum_{i=1}^{n} 2(y_i - \alpha - \beta x_i)(-1) = 0 \qquad (4.61)$$

$$\frac{\partial \Delta^2}{\partial \beta} = \sum_{i=1}^{n} 2(y_i - \alpha - \beta x_i)(-x_i) = 0 \qquad (4.62)$$

from which we get the least-squares estimates of α and β:

$$\hat{\alpha} = \frac{1}{n}\sum_{i=1}^{n} y_i - \frac{\hat{\beta}}{n}\sum_{i=1}^{n} x_i = \bar{y} - \hat{\beta}\bar{x} \qquad (4.63)$$

$$\hat{\beta} = \frac{\sum_{i=1}^{n} x_i y_i - n\bar{x}\bar{y}}{\sum_{i=1}^{n} x_i^2 - n\bar{x}^2} = \frac{\sum_{i=1}^{n}(x_i - \bar{x})\cdot(y_i - \bar{y})}{\sum_{i=1}^{n}(x_i - \bar{x})^2} \qquad (4.64)$$

Strictly speaking, the regression line $E[Y|X = x] = \hat{\alpha} + \hat{\beta}x$ is valid only over the range of values of x for which the data has been observed.

The dual regression line $E[X|Y = y]$ is in general a different linear equation which intersects $E[Y|X = x]$ at $(\bar{x}, \bar{y})$.

The conditional variance $Var[Y|X = x]$ about the regression line can be estimated as:

$$s_{Y|x}^2 = \frac{1}{n-2}\cdot\sum_{i=1}^{n}(y_i - y_i')^2 = \frac{1}{n-2}\cdot\sum_{i=1}^{n}(y_i - \bar{y})^2 - \hat{\beta}^2\sum_{i=1}^{n}(x_i - \bar{x})^2 = \frac{\Delta^2}{n-2}$$

$$(4.65)$$

The physical effect of the linear regression of Y on X can be measured by the reduction of the original variance of Y,

$$s_{Y|x}^2 = \frac{1}{n-1}\cdot\sum_{i=1}^{n}(y_i - \bar{y})^2 \text{ , obtained from taking into account the}$$

general trend with X:

$$r^2 = \frac{s_Y^2 - s_{Y|x}^2}{s_Y^2} \qquad (4.66)$$

The assumptions of linear model and constancy of variance underlying linear regression are, in fact, inherent properties of populations that are jointly normal. In this case we have:

$$E[Y \mid X = x] = \mu_Y + \rho \frac{\sigma_Y}{\sigma_X}(x - \mu_x) \qquad (4.67)$$

$$Var[Y \mid X = x] = \sigma_Y^2 (1 - \rho^2) \qquad (4.68)$$

where ρ is the correlation coefficient (see definition (4.85) below). Thus, if two variates are jointly normal, the regression of Y on X is linear with constant conditional variance and

$$\beta = \rho \frac{\sigma_y}{\sigma_x} ; \alpha = \mu_Y - \beta \mu_X \qquad (4.69)$$

Therefore, if the underlying populations are jointly normal, it is appropriate to use linear regression.

4.7.2   Regression with non-constant variance

The conditional variance about the regression line, $Var[Y \mid X = x]$ may be a function of the independent (controlled) variable. This is the case when the degree of scatter varies with the different values of the controlled variable. This variation may be expressed as:

$$Var[Y \mid X = x] = \sigma^2 g^2 (x) \qquad (4.70)$$

where $g(x)$ is a predefined function and σ is an unknown constant.

In determining the regression equation $E[Y \mid X = x] = \alpha + \beta x$, it would seem reasonable that data points in regions of small variance should have more "weight" than those in regions of large variance. On this premise, we assign weights inversely proportional to the variance:

$$w_i' = \frac{1}{Var[Y \mid X = x_i]} = \frac{1}{\sigma^2 g^2 (x_i)} \qquad (4.71)$$

Then, the squared error is:

$$\Delta^2 = \sum_{i=1}^{n} w_i'(y_i - \alpha - \beta x_i)^2 \qquad (4.72)$$

from which the least-square estimates of α and β become

$$\hat{\alpha} = \frac{\sum_{i=1}^{n} w_i y_i - \hat{\beta} \sum_{i=1}^{n} w_i x_i}{\sum_{i=1}^{n} w_i} ; \hat{\beta} = \frac{\sum_{i=1}^{n}\left(\sum_{i=1}^{n} w_i y_i x_i\right) - \left(\sum_{i=1}^{n} w_i y_i\right)\left(\sum_{i=1}^{n} w_i x_i\right)}{\sum_{i=1}^{n} w_i \left(\sum_{i=1}^{n} w_i x_i^2\right) - \left(\sum_{i=1}^{n} w_i x_i\right)^2}$$

$$(4.73)$$

where, $w_i = \sigma^2 w_i' = \dfrac{1}{g^2(x_i)}$.

An unbiased estimate of the unknown σ^2 is:

$$s^2 = \frac{\sum_{i=1}^{n} w_i(y_i - \hat{\alpha} - \hat{\beta} x_i)^2}{n-2} \qquad (4.74)$$

and an unbiased estimate of the conditional variance is:

$$s_{Y|x}^2 = s^2 g^2(x) \qquad (4.75)$$

4.7.3   Multiple linear regression

Linear regression analysis for more than two variables is simply a generalization of the previous one for two variables. The assumptions underlying multiple regression analysis are as follows:

1.   The mean value of Y is a linear function of $x_1, x_2, ..., x_m$

$$E[Y | x_1, x_2, ..., x_m] = \beta_0 + \beta_1 x_1 + ... + \beta_m x_m \qquad (4.76)$$

2.   The conditioned variance of Y given $x_1, x_2, ..., x_m$ is constant.

$$Var\left[Y \mid x_1, x_2, ..., x_m\right] = \sigma^2 g^2\left(x_1, x_2, ..., x_m\right) \qquad (4.77)$$

The regression analysis then determines estimates for $\beta_0, \beta_1, ..., \beta_m$ and σ^2 based on a set of observed data $(x_1^i, x_2^i, ..., x_m^i, y_i), i = 1, 2, ..., n$.

The function $E\left[Y \mid x_1, x_2, ..., x_m\right]$ can be written also as

$$E\left[Y \mid x_1, x_2, ..., x_m\right] = \alpha + \beta_1\left(x_1 - \bar{x}_1\right) + ... + \beta_m\left(x_m - \bar{x}_m\right) \qquad (4.78)$$

in which the $\bar{x}_i$'s are the sample means of X_i and α is a readjusted constant.

Restricting to the case of constant conditioned variance we have:

$$\Delta^2 = \sum_{i=1}^{n}\left(y_i - y_i'\right)^2 = \sum_{i=1}^{n}\left[y_i - \alpha - \beta_1\left(x_1^i - \bar{x}_1\right) - ... - \beta_m\left(x_m^i - \bar{x}_m\right)\right]^2$$

$$(4.79)$$

Minimizing Δ^2, we get the following estimate

$$\hat{\alpha} = \frac{1}{n}\sum_{i=1}^{n} y_i = \bar{y} \qquad (4.80)$$

and a set of m linear equations involving the m unknowns $\hat{\beta}_0, \hat{\beta}_1, ..., \hat{\beta}_m$:

$$\hat{\beta}_1\sum_{i=1}^{n}\left(x_1^i - \bar{x}_1\right)^2 + \hat{\beta}_2\sum_{i=1}^{n}\left(x_1^i - \bar{x}_1\right)\left(x_2^i - \bar{x}_2\right) + ... + \hat{\beta}_m\sum_{i=1}^{n}\left(x_1^i - \bar{x}_1\right)\left(x_m^i - \bar{x}_m\right) = \sum_{i=1}^{n}\left(x_1^i - \bar{x}_1\right)\left(y_i - \bar{y}\right)$$

$$\vdots$$

$$\hat{\beta}_1\sum_{i=1}^{n}\left(x_m^i - \bar{x}_m\right)\left(x_1^i - \bar{x}_1\right) + \hat{\beta}_2\sum_{i=1}^{n}\left(x_m^i - \bar{x}_m\right)\left(x_2^i - \bar{x}_2\right) + .. + \hat{\beta}_m\sum_{i=1}^{n}\left(x_m^i - \bar{x}_m\right)^2 = \sum_{i=1}^{n}\left(x_m^i - \bar{x}_m\right)\left(y_i - \bar{y}\right)$$

$$(4.81)$$

The conditional variance $Var\left[Y \mid x_1, x_2, ..., x_m\right]$ can be estimated as:

$$s^2_{Y|x_1,\dots,x_m} = \frac{\Delta^2}{n-m-1} = \frac{\sum_{i=1}^{n}\left[y_i - \hat{\alpha} - \hat{\beta}_1\left(x_1^i - \bar{x}_1\right) - \dots - \hat{\beta}_m\left(x_m^i - \bar{x}_m\right)\right]^2}{n-m-1}$$

$$(4.82)$$

4.7.4 Non Linear Regression

Relationships between engineering variables are not always adequately described by linear models. The determination of such non-linear relationships on the basis of observational data involves non-liner regression analysis.

Non-linear regression is usually based on an assumed non-linear mean value function with some unknown coefficients to be evaluated from experimental data.

The simplest type of non-linear regression of Y on x is:

$$E\left[Y\,|\,X = x\right] = \alpha + \beta\,g(x) \qquad (4.83)$$

where $g(x)$ is a predefined non-linear function of x such as $x + x^2$, e^x, lnx.

By defining a new variable $x' = g(x)$,

$$E\left[Y\,|\,X = x'\right] = \alpha + \beta\,x' \qquad (4.84)$$

and transforming the data (x_i, y_i) into $(g(x_i), y_i)$ we are back to linear regression.

4.7.5 Correlation Analysis

The study of the degree of linear interrelation between random variables is called *correlation analysis*. Indeed, the accuracy of a linear model between variables depends on the correlation between them, measured by the so called correlation coefficient,

$$\rho = \frac{Covariance[X,Y]}{\sigma_X \sigma_Y} = \frac{E[(X - \mu_x)(Y - \mu_y)]}{\sigma_X \sigma_Y} \qquad (4.85)$$

Based on a set of observed values of X and Y, the correlation coefficient may be estimated by

$$\hat{\rho} = \frac{1}{n-1} \cdot \frac{\sum_{i=1}^{n}(x_i - \bar{x})(y_i - \bar{y})}{s_X s_Y} = \frac{1}{n-1} \cdot \frac{\sum_{i=1}^{n} x_i y_i - n\bar{x}\bar{y}}{s_X s_Y} \qquad -1 \le \hat{\rho} \le 1 \qquad (4.86)$$

If $\hat{\rho} \approx \pm 1$, then there is strong linear relationship between X and Y, and linear regression analysis is adequate. On the other hand, if $\hat{\rho} \approx 0$, this would indicate a lack of linear relationship between the variables

It is possible to show that,

$$\hat{\rho} = \frac{\sum_{i=1}^{n}(x_i - \bar{x})(y_i - \bar{y})}{\sum_{i=1}^{n}(x_i - \bar{x})^2} \cdot \frac{s_X}{s_Y} = \hat{\beta}\frac{s_X}{s_Y} \qquad (4.87)$$

Furthermore,

$$Var[Y \mid x] = \frac{1}{n-2}\left[\sum_{i=1}^{n}(y_i - \bar{y})^2 - \hat{\rho}^2 \frac{s_Y^2}{s_X^2} \sum_{i=1}^{n}(x_i - \bar{x})^2 \right] = \frac{n-1}{n-2} s_Y^2 (1 - \hat{\rho}^2) \qquad (4.88)$$

from which,

$$\hat{\rho}^2 = 1 - \frac{n-2}{n-1} \cdot \frac{s_{Y|x}^2}{s_Y^2} \xrightarrow{n \to \infty} r^2 = \frac{s_Y^2 - s_{Y|x}^2}{s_Y^2} \qquad (4.89)$$

Thus, we can say that the larger the value of $|\hat{\rho}|$, the greater will be the reduction in the variance when the trend between the variables is taken into account and more accurate will be the prediction based on the regression equation.

Example 4.11 (constant variance) [1]

Tabulated in the first three columns of Table 4.1 are values of shear strengths, in kips per square foot (ksf), obtained from 10 specimens taken at various depths of a clay stratum. Determine the mean and variance of the shear strength as a linear function of depth. Assume that the variance is constant with depth.

Solution

Table 4.1 summarizes the computations in the regression analysis.

Table 4.1: Computational Tableau for the Example

Specimen no.	To determine $\hat{\alpha}$ and $\hat{\beta}$						To determine $s_{Y|x}$	
	Depth (ft) x_i	Strength (ksf) y_i	$x_i y_i$	x_i^2	y_i^2	$y_i' = \alpha + \beta x_i$	$y_i - y_i'$	$(y_i - y_i')^2$
1	6	0.28	1.68	36	0.078	0.325	-0.045	0.0020
2	8	0.58	4.64	64	0.336	0.429	0.151	0.0228
3	14	0.50	7.00	196	0.250	0.739	-0.239	0.0571
4	14	0.83	11.63	196	0.689	0.739	0.091	0.0083
5	18	0.71	12.78	324	0.504	0.946	-0.236	0.0557
6	20	1.01	20.20	400	1.020	1.049	-0.039	0.0015
7	20	1.29	25.80	400	1.662	1.049	0.241	0.0580
8	24	1.50	36.00	576	2.250	1.257	0.243	0.0590
9	28	1.29	36.10	784	1.662	1.463	-0.173	0.0299
10	30	1.58	47.40	900	2.495	1.566	0.014	0.0002
$\sum$	182	9.57	203.23	3876	10.946		$\Delta^2 = 0.2945$	

$$\bar{x} = 18.2$$

$$\bar{y} = 0.957$$

$$\hat{\beta} = \frac{203.23 - 10 \cdot 18.2 \cdot 0.957}{3876 - 10 \cdot 18.2^2} = 0.0516$$

$$\hat{\alpha} = 0.957 - 0.0516 \cdot 18.2 = 0.018$$

$$s_Y^2 = \frac{10.946 - 10 \cdot 0.957^2}{9} = 0.197$$

$$s_{Y|x}^2 = \frac{0.2945}{10 - 2} = 0.0368$$

$$s_{Y|x} = \sqrt{0.0368} = 0.192$$

$$r^2 = 1 - \frac{0.368}{0.197} = 0.813$$

On the basis of the calculations in Table 4.1, the least-squares mean shear strength (in ksf) as a function of depth x is given by

$$E[Y|x] = \alpha + \beta x = 0.018 + 0.0517x \qquad (4.90)$$

whereas the variance of the shear strength at a given depth is estimated to be 0.0368 (ksf)^2, giving $s_{Y|x}^2 = 0.192ksf$. If the linear trend with depth is not taken into account, the unconditional variance of the shear strength would be 0.197 (ksf)^2, and $s_Y = 0.44ksf$. Hence the conditional standard deviation $s_{Y|x}$ is considerably smaller than s_Y.

The regression equation obtained above may be used to predict the shear strength from 6 ft to 30 ft deep. It may not apply to depths beyond 30 ft, unless the linear trend can be justified beyond this depth on physical ground (for example, the same soil type).

Graphically, the regression line obtained above is shown in Fig. 4.17; also shown is the envelope with $\pm s_{Y|x}$ from the regression line.

This represents a band width of one (conditional) standard deviation from either side of the regression line.

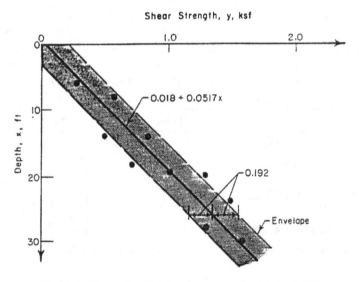

Fig. 4.17: Regression line for shear strength with depth [1]

References

[1] Ang, A.H. and Tang, W.H., Probability Concepts in Engineering Planning and Design. Vol. 1: Basic Principles, 1975.
[2] Zeng, S. W., Reliability Engineering and System Safety, Vol. 55, pp. 151-162, 1997.

5

Reliability of Simple Systems

5.1 Simple system configurations

We consider a system comprised of a set of N independent components, $i = 1, 2, ... N$, each of which has probability p_i of being functioning and $q_i = 1 - p_i$ of being failed. Knowing the probability values p_i, $i = 1, 2, ... N$, and the system configuration, we wish to calculate the probability P that the system is functioning properly.

For time dependent situations we can calculate the reliability of the system $R(t)$ as a function of the components' reliabilities $R_i(t)$, $i = 1, 2, ... N$ [1], [2], [3], [4], [5]. In this case, we may also calculate the mean time to failure, m:

$$m = \int_0^\infty R(t)dt = \tilde{R}(0) \tag{5.1}$$

where $\tilde{R}(s) = \int_0^\infty e^{-st} R(t)dt = L[R(t)]$ is the Laplace transform of $R(t)$.

or

$$m = \underbrace{\int_0^\infty t f(t)dt}_{-dR(t)} = -\frac{d\tilde{f}(s)}{ds}\bigg|_{s=0} \tag{5.2}$$

where $\tilde{f}(s) = L[f(t)]$ and $-\dfrac{d\tilde{f}(s)}{ds} = L[tf(t)]$.

5.2   Series system

Consider the series system of Fig. 5.1. The logic of operation is that all components must function for the system to function.

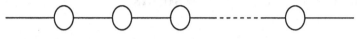

Fig. 5.1: Series System

In terms of the probability that the system functions (intersection of the events that all components function), we have:

$$P = \prod_{i=1}^{N} p_i \tag{5.3}$$

and of the system reliability,

$$R(t) = \prod_{i=1}^{N} R_i(t) \tag{5.4}$$

For exponential components, the system reliability becomes $R(t) = e^{-\lambda t} < R_i(t) = e^{-\lambda_i t}$, i.e. less than the reliability of the less reliable unit, with

$$\lambda = \sum_{i=1}^{N} \lambda_i = \text{system failure rate} \tag{5.5}$$

$$m = \frac{1}{\lambda} = \text{mean time to system failure}$$

The series system is the only logic configuration in which components with constant failure rates induce a constant failure rate for the system. In all other configurations, the reliability of the system is not exponential.

The system fails at $\min(t_1, t_2, ..., t_N)$, where t_i is the failure time of component i.

5.3 Parallel system

Consider the parallel system of Fig. 5.2.

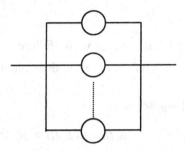

Fig. 5.2: Parallel system: all components perform the same function so that anyone can successfully continue the operation

In terms of probability of the system functioning (union of the individual events of the components functioning) we have:

$$P = 1 - \prod_{i=1}^{N} (1 - p_i) \qquad (5.6)$$

which in terms of reliability becomes:

$$R(t) = 1 - \prod_{i=1}^{N} [1 - R_i(t)] \qquad (5.7)$$

For N exponential components with different failure rates,

$$R(t) = 1 - \prod_{i=1}^{N} [1 - e^{-\lambda_i t}] \qquad (5.8)$$

$$m = \sum_{i=1}^{N} \frac{1}{\lambda_i} - \sum_{i=1}^{N-1} \sum_{j=i+1}^{N} \frac{1}{[\lambda_i + \lambda_j]} + \sum_{i=1}^{N-2} \sum_{j=i+1}^{N-1} \sum_{k=j+1}^{N} \frac{1}{[\lambda_i + \lambda_j + \lambda_k]} - \cdots + (-1)^{N-1} \frac{1}{\sum_{i=1}^{N} \lambda_i}$$

$$(5.9)$$

Since the system fails when all its elements fail, the time-to-failure of the system is $max(t_1, t_2, ..., t_N)$

Example 5.1

Consider two exponential units, with failure rates λ_1 and λ_2, respectively. The system reliability is time-dependent

$$R(t) = 1 - (1 - e^{-\lambda_1 t})(1 - e^{-\lambda_2 t}) =$$

$$= \underbrace{e^{-\lambda_1 t}}_{R_1} + \underbrace{e^{-\lambda_2 t}}_{R_2} - e^{-(\lambda_1 + \lambda_2)t} > R_1(t) \text{ and } R_2(t) > e^{-\lambda t} = e^{-(\lambda_1 + \lambda_2)t} \quad \text{(series)}$$

$$m = \frac{1}{\lambda_1} + \frac{1}{\lambda_2} - \frac{1}{[\lambda_1 + \lambda_2]}$$

In the case of identical elements, we can compare the series and parallel configurations:

$$
\left.
\begin{array}{ll}
\text{parallel} & m = \sum_{n=1}^{N} \frac{1}{n\lambda} \\[2em]
\text{series} & m = \dfrac{1}{N\lambda}
\end{array}
\right\}
\Rightarrow
\lambda m_{series} = \frac{1}{N} < \sum_{n=1}^{N} \frac{1}{n} = \lambda m_{parallel}
$$

5.4   *r*-out-of-*N* systems

Consider N identical components which function in parallel but only $r < N$ are needed for the system to function (the parallel system is a particular case with $r = 1$). In terms of probability of the system functioning:

$$P\left\{any\ k\ of\ N\ functioning\right\} = P_k = \binom{N}{k} p^k (1 - p)^{N-k} \quad \text{(binomial)}$$

$$(5.10)$$

$$P\left\{at\ least\ r\ of\ N\ functioning\right\} = P = \sum_{k=r}^{N} P_k = \sum_{k=r}^{N} \binom{N}{k} p^k (1-p)^{N-k}$$

$$(5.11)$$

Considering N exponential components with equal λ, the system reliability reads:

$$R(t) = \sum_{k=r}^{N} \binom{N}{k} e^{-\lambda k t} \left(1 - e^{-\lambda t}\right)^{N-k} \tag{5.12}$$

with mean time to failure equal to:

$$m = \sum_{k=r}^{N} \frac{1}{k\lambda}$$

which gives the mean time to $r+1$ failures, given that r successes are required for the system success.

5.5   Standby systems

A common feature of the previous series and parallel systems is that they are not time dependent: the logic of the system dictates the eventual time dependence so that the expressions for reliability are derived by simply replacing p_i with $R_i(t)$. The formulas for P may be interpreted as holding at every point in time and the state of the system is determined by the present state of its components. This is no longer true for standby systems for which the whole story of the system from $t = 0$ must be considered.

In a standby system, one component is functioning and when it fails it is replaced immediately by another component (sequential operation of one component at a time) (Fig. 5.3).

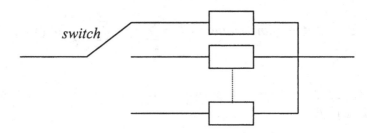

Fig. 5.3: Standby system

5.5.1   Cold Standby

A cold standby is a configuration in which the standby component is not subject to failure until it is switched on. Moreover, the switch is assumed to be perfect.

Since the components are operated sequentially, the system fails at time $T = \sum_{i=1}^{N} T_i$, which is a random variable sum of N independent random variables. The pdf of T can then be found with the use of the convolution theorem.

For simplicity, consider a nominal component 1 and a standby component 2, with random failure times T_1 and T_2 distributed as $f_1(t)$ and $f_2(t)$, respectively. The probability density function $f_T(t)$ of the system failure $T = T_1 + T_2$ is given by the convolution product:

$$f_T(t) = f_1(t) * f_2(t) = \int_0^t f_1(x) f_2(t-x) dx = f_1(t) * f_2(t) \qquad (5.13)$$

Where the symbol * indicates the convolution product. Taking the Laplace transform,

$$\tilde{f}_T(s) = L[f_1(t) * f_2(t)] = \tilde{f}_1(s) \tilde{f}_2(s)$$

Generalizing, for a system with 1 nominal component and N-1 standby components we have,

$$\tilde{f}_T(s) = \prod_{i=1}^{N} \tilde{f}_i(s) \tag{5.14}$$

Inverse-transforming $\tilde{f}_T(s)$ into $f_T(t)$, we can obtain the system reliability

$$R(t) = 1 - \int_0^t f_T(x)dx \tag{5.15}$$

Example 5.2

Consider N identical exponential components, all with failure rate λ. The probability density function of the components times to failure and corresponding Laplace transforms are:

$$f_i(t) = \lambda e^{-\lambda t} \qquad\qquad \tilde{f}_i(s) = \frac{\lambda}{s+\lambda}, \; i=1,2,...N$$

From (5.14), the system Laplace transform $\tilde{f}_T(s)$ is:

$$\tilde{f}_T(s) = \frac{\lambda^N}{(s+\lambda)^N}$$

Inverse-transforming

$$f_T(t) = \frac{\lambda^N t^{N-1}}{(N-1)!} e^{-\lambda t} \qquad \text{(Gamma distribution)}$$

and from (5.15):

$$R(t) = e^{-\lambda t} \sum_{k=0}^{N-1} \frac{(\lambda t)^k}{k!} \qquad \text{(Poisson distribution)}$$

The system reliability shows that with $N-1$ cold standby components, the system can sustain up to $N-1$ failures (including that of the operating component as well) and still be functioning: The mean time to sytem failure m is:

$$m = -\left.\frac{d\tilde{f}_T(s)}{ds}\right|_0 = \frac{N}{\lambda}$$

Example 5.3

Consider two different exponential components with failure rates λ_1 and λ_2, respectively.

From   (5.14), the system Laplace transform $\tilde{f}_T(s)$ is:

$$\tilde{f}_T(s) = \frac{\lambda_1}{s+\lambda_1}\frac{\lambda_2}{s+\lambda_2}$$

Inverse-transforming,

$$f_T(t) = \frac{\lambda_1\lambda_2}{\lambda_1-\lambda_2}\left(e^{-\lambda_2 t} - e^{-\lambda_1 t}\right)$$

and from   (5.15):

$$R(t) = \frac{\lambda_1 e^{-\lambda_2 t} - \lambda_2 e^{-\lambda_1 t}}{\lambda_1 - \lambda_2}$$

$$m = \frac{1}{\lambda_1} + \frac{1}{\lambda_2}$$

For N dissimilar exponential components, with failure rates λ_i, $i = 1,2,...N$, the mean time to failure becomes:

$$m = \sum_{i=1}^{N} \frac{1}{\lambda_i}$$

Note that the purpose of standby units is to increase the reliability and the system MTTF, m, over the values which would be obtained without it. Indeed, comparing with the parallel configuration:

$$\lambda\, m_{\text{parallel}} = \sum_{n=1}^{N} \frac{1}{n} < N = \lambda\, m_{\text{standby}}$$

Example 5.4 (Imperfect switching)

Consider the case of two different exponential components 1 and 2 with failure rates λ_1 and λ_2, respectively, and component 2 in cold standby, and an imperfect switch with constant probability of good switching equal to R_{sw}. There are two mutually exclusive ways for the system to survive up to t:

i)   Switch fails: reliability of the system $R(t) = e^{-\lambda_1 t}$ (the system can rely only on component 1).
ii) Switch does not fail: reliability of the system = reliability of the system as if the switch were perfect:

$$R(t) = \frac{\lambda_1 e^{-\lambda_2 t} - \lambda_2 e^{-\lambda_1 t}}{\lambda_1 - \lambda_2}$$

Then, the system reliability is:

$$R(t) = (1 - R_{sw}) e^{-\lambda_1 t} + R_{sw} \frac{\lambda_1 e^{-\lambda_2 t} - \lambda_2 e^{-\lambda_1 t}}{\lambda_1 - \lambda_2}$$

5.5.2   Hot Standby

Up to now we were able to assume independent failures: the failure of any unit was not influenced by the failures of the other units. In the case of a hot standby this is not so since the standby unit has a finite probability to fail also while in standby.

Let $f_1(t)$ be the pdf of the time to failure of the component, $f_s(t)$ that of the standby unit while in standby and $f_2(t)$ that of the standby unit when online. Let $R_1(t), R_s(t)$ and $R_2(t)$ be the corresponding reliabilities.

The convolution theorem can no longer be used to calculate the reliability of the system, because there is no independence of the failure events any more.

The system will perform its task in the interval $(0, t)$ in either of two mutually exclusive ways:

(i)   the online component 1 does not fail in $(0, t)$, with probability

$$R_1(t) = 1 - \int_0^t f_1(x)dx ;$$

(ii) the online component fails in $(\tau, \tau + d\tau)$, with probability
      $f_1(\tau)d\tau$;

      the standby component 2 does not fail in $(0, \tau)$, with probability $R_s(\tau)$ and it operates successfully from τ to t with probability $R_2(t-\tau)$.

Then, the system reliability is given by the sum of the probabilities of the two mutually exclusive events:

$$R(t) = R_1(t) + \int_0^t f_1(\tau)d\tau R_s(\tau)R_2(t-\tau)   \qquad (5.16)$$

For exponential components,

$$R(t) = e^{-\lambda_1 t} + \int_0^t \lambda_1 e^{-\lambda_1 \tau} e^{-\lambda_s \tau} e^{-\lambda_2(t-\tau)} d\tau =$$

$$= e^{-\lambda_1 t} + \frac{\lambda_1}{\lambda_1 + \lambda_s - \lambda_2}\left[e^{-\lambda_2 t} - e^{-(\lambda_1 + \lambda_s)t}\right]   \qquad (5.17)$$

In the limit cases:

$$\lambda_s = 0 \qquad \rightarrow \quad R(t) = e^{-\lambda t}(1 + \lambda t) \qquad\qquad \text{cold standby}$$

$$\left.\begin{array}{l} \lambda_1 = \lambda_2 = \lambda \\ \lambda_s = \lambda \end{array}\right\} \quad \rightarrow \quad R(t) = 2e^{-\lambda t} - e^{-2\lambda t}$$

parallel

References

[1] Rausand, M. and Hoyland, A., System Reliability Theory, Wiley, 2004.
[2] Ushakov, I.A., Handbook of Reliability Engineering, Wiley, 1994.
[3] Schneeweiss, W.G., Reliability Modeling, LiLoLe-Verlag, 2001.
[4] Birolini, A., Reliability Engineering, Springer, 2004.
[5] Lewis, E.E., Introduction to Reliability Engineering, Wiley, 1996.

6

Availability and Maintainability

6.1 Introduction

Reliability and availability represent important performance parameters of a system, with respect to its ability to fulfil the required mission during a given functioning period [1], [2], [3], [4], [5]. From this point of view, two main types of systems can be defined:

1. Systems which must satisfy a specified mission within an assigned period of time: in this case, the reliability is the appropriate performance indicator of the ability to achieve the desired objective without failures;
2. Systems maintained: in this case, the availability quantifies in a suitable way the system ability to fulfill the assigned mission at any specific moment of its life time. Basic maintenance procedures can be distinguished in:
   a. Off-schedule (corrective): this amounts to the replacement or repair of failed units;
   b. Preventive: this amounts to performing regular inspections, and possibly repair, following a given maintenance plan;
   c. Conditioned: it amounts to performing a repair action upon detection of degradation.

6.2 Availability definition

As above said, an important figure of merit for a system undergoing maintenance (i.e., corrective, preventive or conditioned maintenance) is its (un)availability [1], [2], [3], [4], [5].

Let $X(t)$ be an indicator variable denoting the state at time t of a system undergoing maintenance, such that (Fig. 6.1):

$X(t)=1$, system is operating at time t

$X(t)=0$, system is failed at time t

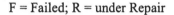

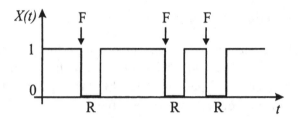

Fig. 6.1: System state indicator variable

The *instantaneous availability p(t)* and *unavailability q(t)* are defined as the probability that the system is operating at time t and as the probability that the system is failed at time t, respectively:

$$p(t) = P[X(t) = 1] = E[X(t)] \qquad (6.1)$$

$$q(t) = P[X(t) = 0] = 1 - p(t) \qquad (6.2)$$

Notice the difference in the meaning of *p(t)*, the probability that the system is functioning at time t, from the reliability $R(t)$, i.e. the probability that the system functions continuously with no failures up to time t.

To judge the performance of a maintainable system, so as to be able to compare different maintenance strategies, we need to define appropriate quantities for an average description of its probabilistic behavior. We distinguish two cases:

−   For components whose behavior can be described by finite Markov processes, we introduce the *limiting* or *steady state availability*:

$$p = \lim_{t \to \infty} p(t) \qquad (6.3)$$

By definition, it represents the probability that the component is functioning at an arbitrary moment of time, after the transient failure and repair processes have stabilized. It is obviously an undefined measure for systems under periodic maintenance, for which the limit does not exist.

- For components under periodic maintenance, the average availability over a given period of time T is introduced as the proper indicator of system performance, and it is given by:

$$p_T = \frac{1}{T} \cdot \int_0^T p(t)dt = \frac{\overline{UP\,time}}{T} \qquad (6.4)$$

where $\overline{UPtime}$ is the average time the system is functioning (UP) within T. From the definition, it follows that p_T is not a probability, but represents the expected proportion of time that the system is operating in $[0, T]$. At steady-state, the limiting average availability can be defined as:

$$p_\infty = \lim_{T \to \infty} \frac{1}{T} \cdot \int_0^T p(t)dt \qquad (6.5)$$

Note that if the limiting availability p exists, then $p_\infty = p$.

6.3 Contributions to unavailability

The main contributions to the unavailability of a system generally come from:

1. *Unrevealed failure*, i.e. when a stand-by component fails unnoticed. The system goes on without noticing the component failure until a test on the component is made or the component is demanded to function.

2.   *Testing/preventive maintenance,* i.e. when a component is removed
      from the system because
      it has to be tested or must undergo preventive maintenance.
3.   *Repair,* i.e. when a component is unavailable because under repair.

6.4   The availability of an unattended component (no repairs)

An unattended component will function till its first failure and remain
failed after that, since repairs are not allowed. Hence, the probability *q(t)*
that at time *t* the component is not functioning is equal to the probability
that it failed before *t*, i.e. the cumulative failure probability *F(t)*. In other
words, the instantaneous unavailability of the component will be equal to
the cumulative distribution function of failure times:

$$q(t) \equiv F(t) \tag{6.6}$$

and the component availability will be equal to its reliability:

$$p(t) = 1 - q(t) \equiv R(t) \tag{6.7}$$

6.5   The availability of a continuously monitored component

For a continuously monitored component it is assumed that restoration
starts immediately after its failure. Still, we need to define the
probabilistic model describing the duration of the repair process.

    We indicate with *G(t)* the cumulative distribution function of the
random time duration of the repair process:

$$G(t)=P\{repair\ process\ ends\ before\ t\ units\ from\ failure\} \tag{6.8}$$

and with *g(t)* the corresponding probability density function.

    To analyze the failure and repair processes, we suppose that we start
with *N* items at time *t* = 0. At any successive time *t*, some items will be
functioning (UP) whereas the others will be failed (DOWN), so that the
total number *N* is conserved.

We can then establish a balance equation between time t and time $t+\Delta t$. At time t, the number of items which are UP is $N \cdot p(t)$; at time $t+\Delta t$ the number of items which are UP is $N \cdot p(t + \Delta t)$.

Assuming, for simplicity, that the components have exponentially distributed failure times with rate λ, then $\lambda \cdot \Delta t$ is the conditional failure probability in Δt, given that the item was UP at time t. Considering that $p(t)$ is the probability of the item being UP at time t, at the beginning of Δt, we get the unconditional failure probability $p(t) \cdot \lambda \cdot \Delta t$. Thus, the number of items failing during the interval Δt, i.e. the loss term in the balance equation is given by $N \cdot p(t) \cdot \lambda \cdot \Delta t$.

Following the same logic, we obtain the gain term of the balance equation due to components that had failed in $(\tau, \tau + \Delta \tau)$ and whose restoration terminates in $(t, t+\Delta t)$ (Fig. 6.2).

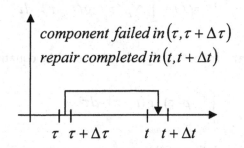

Fig. 6.2: Gain term due to restoration of components in Δt

Obviously, the failure can occur at any $\tau \le t$ so that we need to integrate over time:

$$\int_{0}^{t} N \cdot p(\tau) \cdot \lambda \cdot \Delta \tau \cdot g(t - \tau) \cdot \Delta t \qquad (6.9)$$

where,

$p(\tau) \cdot \lambda \cdot \Delta \tau$ is the item unconditional probability of failing in the interval $\Delta \tau$ ($\lambda \cdot \Delta \tau$ is the conditional failure probability in

$\Delta\tau$), knowing that the item was UP at time τ ($p(\tau)$ is the probability of the *i-th* item being UP at time τ).

$g(t-\tau)\cdot\Delta t$ is the probability of completing in (t, $t+\Delta t$) the restoration which had started upon failure in (τ, $\tau+\Delta\tau$).

The balance equation then writes:

$$N \cdot p(t+\Delta t) = N \cdot p(t) - N \cdot p(t) \cdot \lambda \cdot \Delta t + \int_0^t N \cdot p(\tau) \cdot \lambda \cdot \Delta\tau \cdot g(t-\tau) \cdot \Delta t$$

$$(6.10)$$

Dividing by $N\cdot\Delta t$, subtracting $p(t)$ on both sides and letting Δt tend to zero, we obtain the integral-differential form of the balance:

$$\frac{dp(t)}{dt} = -\lambda \cdot p(t) + \int_0^t \lambda \cdot p(\tau) \cdot g(t-\tau) \cdot d\tau \qquad (6.11)$$

where the integral term on the right-hand side of the equation,

$$\int_0^t \lambda \cdot p(\tau) \cdot g(t-\tau) \cdot d\tau \qquad (6.12)$$

represents the convolution of the instantaneous availability function and the restoration probability density function.

As initial condition of the integral-differential equation (6.11), we will assume, in general, $p(0)=1$, which means that the component is UP at the initial time.

The solution to the integral-differential equation (6.11) can be easily obtained introducing the Laplace transforms:

$$f(x) \quad : \qquad L[f(x)] = \tilde{f}(s) = \int_0^\infty e^{-s\cdot x} f(x)dx$$

$$(6.13)$$

$$\frac{df(x)}{dx} \quad : \qquad L\left[\frac{df(x)}{dx}\right] = s \cdot \tilde{f}(s) - f(0)$$

Applying the Laplace transform to the balance equation (6.11) we obtain the following algebraic equation in the unknown $\tilde{p}(s)$:

$$s \cdot \tilde{p}(s) - 1 = -\lambda \cdot \tilde{p}(s) + \lambda \cdot \tilde{p}(s) \cdot \tilde{g}(s) \qquad (6.14)$$

which can be solved for $\tilde{p}(s)$:

$$\tilde{p}(s) = \frac{1}{s + \lambda \cdot \left(1 - \tilde{g}(s)\right)} \qquad (6.15)$$

Applying the inverse Laplace transform to $\tilde{p}(s)$, the instantaneous availability $p(t)$ is determined.

Furthermore, to determine the limiting availability, p_∞, the final-value theorem can be exploited:

$$p_\infty = \lim_{t \to \infty} p(t) = \lim_{s \to 0}\left[s \cdot \tilde{p}(s)\right] = \lim_{s \to 0}\left[\frac{s}{s + \lambda \cdot \left(1 - \tilde{g}(s)\right)}\right] \qquad (6.16)$$

As s tends to 0, a first order approximation of $\tilde{g}(s)$ can be considered:

$$\tilde{g}(s) = \int_0^\infty e^{-s\tau} g(\tau) d\tau = \int_0^\infty (1 - s \cdot \tau + \ldots) g(\tau) d\tau \cong 1 - s \cdot \int_0^\infty \tau \cdot g(\tau) d\tau = 1 - s \cdot \bar{\tau}_R$$

$$(6.17)$$

where $\bar{\tau}_R$ is the expected value of the restoration time distribution $G(t)$, also called the mean-time-to-repair, MTTR.

Hence,

$$p_\infty = \lim_{s \to 0} \frac{s}{s + \lambda \cdot s \cdot \bar{\tau}_R} = \frac{1}{1 + \lambda \cdot \bar{\tau}_R} = \frac{1/\lambda}{1/\lambda + \bar{\tau}_R} =$$

$$= \frac{MTTF}{MTTF + MTTR} = \frac{average\ time\ the\ component\ is\ UP}{average\ period\ of\ a\ failure\ /\ repair\ "cycle"}$$

$$(6.18)$$

Note that this result is valid for any repair process $G(t)$.

Example 6.1

Find the instantaneous and the limiting availabilites for a component whose restoration probability density is:

$$g(t) = \mu \cdot e^{-\mu \cdot t}$$

Solution:

The Laplace transform of the restoration density is:

$$\tilde{g}(s) = L[g(t)] = \frac{\mu}{s + \mu}$$

Then, substituting $\tilde{g}(s)$ in the above expression (6.15) for $\tilde{p}(s)$, we get:

$$\tilde{p}(s) = \frac{1}{s + \lambda \cdot \dfrac{s}{s + \mu}} = \frac{s + \mu}{s \cdot (s + \mu + \lambda)}$$

Applying the inverse Laplace transform, we obtain the instantaneous availability:

$$p(t) = \frac{\mu}{\mu + \lambda} + \frac{\lambda}{\mu + \lambda} \cdot e^{-(\mu + \lambda)t}$$

and the limiting availability is:

$$P_\infty = \frac{\mu}{\mu + \lambda}$$

which can also obtained directly from (6.18).

6.6 The availability of a component under periodic test and maintenance

Safety systems are generally in standby until an accident occurs, which calls for their operation. Hence, their components must be periodically tested. The components are unattended between tests and their failure is revealed only when tested.

For a component under periodic test and maintenance, the instantaneous unavailability is a periodic function of time, and, as such, it does not posses a limit. In this case, the performance indicator used is the average unavailability. The calculation of the average unavailability over a period of time T utilizes its definition:

$$q_T = \frac{1}{T} \cdot \int_0^T q(t)\,dt = \frac{\overline{DOWNtime}}{T} \qquad (6.19)$$

where $\overline{DOWNtime}$ is the average time the system is failed (DOWN) within T. For simplicity, let us consider the simple case of the unavailability being due to unrevealed random failures that can occur at any moment of time with constant rate λ.

Assuming instantaneous and perfect testing and maintenance procedures, the instantaneous availability within a period τ coincides with the reliability because the component is unattended between two successive maintenance times, i.e. between $(k-1)\tau$ and $k\tau$, k=1, 2, ...

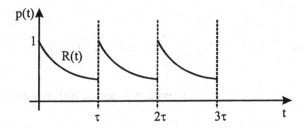

Fig. 6.3: Availability of a component under periodic test and maintenance, with period τ

Note that since $p(t)$ and $R(t)$ are periodic functions we can derive them within one period.

For the calculation of the average unavailability, we refer to Fig. 6.4 which shows a generic random behavior of the component under periodic test and maintenance.

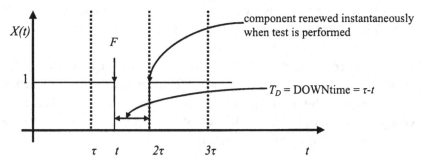

Fig. 6.4: State indicator variable for a component under periodic test and maintenance

The average unavailability within one period τ is, by definition:

$$q_\tau = \frac{\overline{DOWNtime}}{\tau} = \frac{\overline{T}_D}{\tau} \tag{6.20}$$

where the mean DOWNtime, $\overline{T}_D$, is:

$$\overline{T}_D = \int_0^\tau (\tau - t) f(t) dt = \int_0^\tau (\tau - t) dF \tag{6.21}$$

Integrating by parts:

$$\overline{T}_D = (\tau - t) \cdot F(t) \Big|_0^\tau + \int_0^\tau F(t) dt = \int_0^\tau F(t) dt \tag{6.22}$$

Hence, the average unavailability within τ has the following expression:

$$q_\tau = \frac{\overline{T}_D}{\tau} = \frac{\int_0^\tau F(t) dt}{\tau} \tag{6.23}$$

and the average availability:

$$p_\tau = \frac{\overline{T}_U}{\tau} = \frac{\int_0^\tau R(t)dt}{\tau} \tag{6.24}$$

where $\overline{T}_U$ is the mean UPtime within τ.

Expressions (6.23) and (6.24) are just the definitions of the average unavailability and availability over the period τ, since $q(t) = F(t)$ and $p(t) = R(t)$ within the interval τ in which the component is unattended. Then, we are in the situation that for different systems, with different logics of redundancy, we can compute q_τ, p_τ by first computing their failure distribution and reliability, according to the logic of operation, and then applying the above expressions.

If the component has exponentially distributed failure times with constant rate λ, we have that the cumulative distribution is:

$$F(t) = 1 - e^{-\lambda t} \tag{6.25}$$

For failure rates and for times such that the inequality $\lambda \cdot t \leq 0.10$ is satisfied, then the cumulative distribution function could be approximated as follows:

$$F(t) = 1 - e^{-\lambda t} \cong \lambda \cdot t \tag{6.26}$$

and the average unavailability would take the form:

$$q_\tau = \frac{\int_0^\tau F(t)dt}{\tau} = \frac{\int_0^\tau \lambda \cdot t \, dt}{\tau} = \frac{1}{2} \cdot \lambda \tau \tag{6.27}$$

Intuitively, we would expect the component with constant failure rate to fail halfway the period.

Finally, assuming a finite repair time τ_R, this must be counted as DOWNtime, if significant. Hence, the average unavailability and availability over the complete maintenance cycle period $\tau + \tau_R$ will change into:

$$\overline{q} = \frac{\tau_R + \int\limits_0^\tau F(t)\,dt}{\tau + \tau_R} \qquad (6.28)$$

$$\overline{p} = \frac{\int\limits_0^\tau R(t)\,dt}{\tau + \tau_R} \qquad (6.29)$$

If the repair time τ_R is small compared with the period τ, we get:

$$\overline{q} = \frac{\tau_R + \int\limits_0^\tau F(t)\,dt}{\tau} \qquad (6.30)$$

$$\overline{p} = \frac{\int\limits_0^\tau R(t)\,dt}{\tau} \qquad (6.31)$$

6.6.1 Single component under periodic maintenance: a more realistic case

To compute the average unavailability of a component over its lifetime [0, T], we need to compute the average DOWNtime and then compute the average unavailability using its definition:

$$\overline{q}_{0T} = \frac{\overline{T}_{D(0T)}}{T} \qquad (6.32)$$

Before making any prediction on the component unavailability, we must define its failure characteristics, underlying the causes which lead the component into a malfunctioning state. We consider the following causes of failure:

– random failure at any time, modeled by the cumulative distribution function $F(t)$
– on-line switching failure on demand, with occurrence probability Q_0
– maintenance disabling the component, with probability γ_0 (due to human error during inspection, testing or repair)

An example of the latter cause could be forgetting to return a manually operated valve to proper configuration after testing (typical occurrence probability, $\gamma_0 = 10^{-2}$).

Let us assume that the component is initially working i.e., $q(0) = 0$; $p(0) = 1$. In order to compute the component average unavailability $\overline{q}_{0T}$, we refer to its timeline of Fig. 6.5.

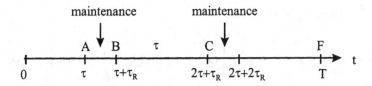

τ = time period between successive maintenances
τ_R = duration of a maintenance action
T = component lifetime

Fig. 6.5: Timeline of a component under periodic maintenance

We have:

• $\overline{0A}$ – From the initial working state at time 0 to the first maintenance (A), the probability of finding the component DOWN at the generic time t is due either to the fact that it was demanded to start but failed or to the fact that it randomly failed unrevealed before t. Thus, the instantaneous unavailability at t, $0 < t < \tau$, reads:

$$q_{0A}(t) = Q_0 + (1 - Q_0) \cdot F(t) \qquad (6.33)$$

and the average DOWNtime:

$$\overline{T}_{D(0A)} = \int_0^\tau q_{0A}(t)dt = Q_0 \cdot \tau + (1-Q_0) \cdot \int_0^\tau F(t)dt \qquad (6.34)$$

- $\overline{AB}$ – during the maintenance period, the component remains disconnected and, thus, the average DOWNtime is the whole maintenance time:

$$\overline{T}_{D(AB)} = \tau_R \qquad (6.35)$$

- $\overline{BC}$ – at the generic time t between two maintenances, the component can be found failed because, by error, it remained disabled from the previous maintenance or as before, because it failed on demand or randomly before t. Thus the instantaneous unavailability at time t is given by

$$q_{BC}(t) = \gamma_0 + (1-\gamma_0) \cdot \left[Q_0 + (1-Q_0) \cdot F(t)\right] \qquad (6.36)$$

and the average downtime:

$$\overline{T}_{D(BC)} = \gamma_0 \cdot \tau + (1-\gamma_0) \cdot \left[Q_0 \cdot \tau + (1-Q_0) \cdot \int_0^\tau F(t)dt\right] \qquad (6.37)$$

- $\overline{CF}$ – The normal maintenance cycle is repeated throughout the component lifetime T. The number of repetitions, i.e. the number of AB-BC maintenance cycles, is:

$$k = \frac{T}{\tau + \tau_R} \qquad (6.38)$$

Then, the average DOWNtime between the first maintenance occurrence and the end of the lifetime T (thus, excluding the negligible first transient interval to first maintenance, 0A, which is typically much smaller than T) is:

$$\overline{T}_{D(AF)} = \frac{T}{\tau + \tau_R} \cdot \left\{ \tau_R + \gamma_0 \tau + (1-\gamma_0) \cdot \left[Q_0 \cdot \tau + (1-Q_0) \cdot \int_0^\tau F(t)dt \right] \right\}$$

(6.39)

whereas the total expected DOWNtime (including the first transient interval to first maintenance) would be:

$$\overline{T}_{D(0T)} = Q_0 + (1-Q_0) \cdot \int_0^\tau F(t)dt + \frac{T}{\tau + \tau_R} \left\{ \tau_R + \gamma_0 \tau + (1-\gamma_0) \left[Q_0 \cdot \tau + (1-Q_0) \cdot \int_0^\tau F(t)dt \right] \right\}$$

(6.40)

Correspondingly, the average unavailability over the component lifetime T becomes:

$$\overline{q}_{0T} = \frac{\overline{T}_{D(0T)}}{T} = \frac{Q_0}{T} + \frac{1-Q_0}{T} \cdot \int_0^\tau F(t)dt + \frac{1}{\tau + \tau_R} \left\{ \tau_R + \gamma_0 \tau + (1-\gamma_0) \left[Q_0 \cdot \tau + (1-Q_0) \cdot \int_0^\tau F(t)dt \right] \right\}$$

(6.41)

Neglecting the first contribution related to the transient period 0A, because Q_0 and $F(t)$ are generally very small and T is large, since typically $\tau_R \ll \tau$ and $\tau \ll T$, the average unavailability can be realistically simplified to:

$$\overline{q}_{0T} \cong \frac{\tau_R}{\tau} + \gamma_0 + (1-\gamma_0) \cdot \left[Q_0 + \frac{1-Q_0}{\tau} \cdot \int_0^\tau F(t)dt \right]$$

(6.42)

Considering an exponential component with small, constant failure rate λ and, thus, a cumulative distribution function approximated as:

$$F(t) = 1 - e^{-\lambda t} \cong \lambda \cdot t$$

(6.43)

the average unavailability takes the form:

$$\overline{q}_{0T} \cong \frac{\tau_R}{\tau} + \gamma_0 + (1-\gamma_0) \cdot \left[Q_0 + (1-Q_0) \cdot \frac{1}{2} \cdot \lambda \cdot \tau \right]$$

(6.44)

Often in practice, $\gamma_0 \ll 1$, $Q_0 \ll 1$. Then:

$$\overline{q}_{0T} \cong \frac{\tau_R}{\tau} + \gamma_0 + Q_0 + \frac{1}{2} \cdot \lambda \cdot \tau \qquad (6.45)$$

From this formula it is possible to distinguish each contribution to the unavailability of the component as follows:

$\dfrac{\tau_R}{\tau}$ \qquad unavailability during maintenance

γ_0 \qquad unavailability due to an error which leaves the unit DOWN after test

Q_0 \qquad unavailability due to the switch failing on demand

$\dfrac{1}{2} \cdot \lambda \cdot \tau$ \qquad unavailability due to random, unrevealed failures between successive tests

6.7   Maintainability

When it is observed that a system or piece of equipment fails to perform its function satisfactorily, all or part of it is taken out of operation to locate and correct the fault. The fault may be corrected by a repair or a part may be replaced by a spare.

When it has been verified by appropriate test that the fault is corrected, the equipment is returned to service. It may be placed back in operation, or it may be placed in standby, depending on the operational conditions at the time.

The total time from system failure until return to service constitutes the system DOWNtime. DOWNtime can be divided into two categories [6]:

a)   active repair time – sensitive to environment, technician skill level, procedures, etc.

b)   administrative time – sensitive to administrative procedures, filing, storage, etc.

Active repair can be divided into recognition or detection time, fault location or diagnosis time, correction or repair time, and verification of final malfunction check time. The administrative contribution to DOWNtime is that required to obtain the spare or in waiting for personnel, manuals, tools, or test and tuning equipment.

The time required to perform the activities associated to each of these categories varies statistically from one failure to another, depending on the conditions associated with the particular maintenance events. The variety of alternate courses of action that maintenance technicians may follow in this repair process suggests both a large number of relatively short-time repair periods and a smaller number of long periods. The former would correspond to the more usual case where the failed unit is replaced by a spare at the operational site upon detection of a failure. The long DOWNtimes would occur when diagnosis is difficult or no spare is immediately available, and might represent the length of time to repair the failure at the maintenance area. This is why in practice the log-normal distribution often is a good representation of maintenance action times.

System maintainability is defined as the probability that an item will be restored to specified conditions within a given period of time when maintenance action is performed in accordance with prescribed procedures and resources [6].

Let T_D denote the item DOWNtime random variable, distributed according to a density function $g(t)$. Then, maintainability can be written as [6]:

$$P(T_D \leq T) = \int_0^T g(t)dt \qquad (6.46)$$

and the mean DOWNtime $\overline{T}_D$ is:

$$\overline{T}_D = \int_0^\infty t \cdot g(t)dt \qquad (6.47)$$

The maintainability analysis of a system is focused on the calculation of $\overline{T}_D$.

6.8   A policy of preventive and corrective maintenance

We present an example of a maintenance policy applied to a component in continuous operation, which encompasses both a corrective maintenance action upon failure and a preventive, periodic maintenance [6].

Let us introduce the following notation for the scheduled, periodic maintenance and the corrective, emergency maintenance upon failure:

τ time of continuing operation without failure, after which we perform the scheduled maintenance; in other words, it represents the maintenance period between two successive maintenances. Note that we allow τ to be infinite, in which case the preventive maintenance is not scheduled;

τ_R time interval required to perform a scheduled maintenance action;

t time of system failure in correspondence of which a corrective, emergency maintenance action is started;

τ_e time interval required to perform the emergency maintenance action.

We assume that any maintenance action restores the system "as good as new".

The quantities of interest are:

–   the Mean Time Between Failures MTBF; it is the item UPtime, i.e. the mean operating time until replacement, which takes into account the two mutually exclusive scenarios of no failure within the period τ and failure at time t within τ.

$$MTBF = \tau \cdot R(\tau) + \int_0^\tau t \cdot f(t)dt = \int_0^\tau R(t)dt \qquad (6.48)$$

－ the DOWNtime $\overline{T}_D$; it is the mean time needed to replace the item at failure or to repair it at the scheduled maintenance:

$$\overline{T}_D = \tau_e \cdot [1 - R(\tau)] + \tau_R \cdot R(\tau) \qquad (6.49)$$

The average availability of the component is then:

$$p_T = \frac{UPtime}{UPtime + DOWNtime} = \frac{MTBF}{MTBF + \overline{T}_D} = \frac{\int_0^\tau R(t)\,dt}{\int_0^\tau R(t)\,dt + \tau_e \cdot [1 - R(\tau)] + \tau_R \cdot R(\tau)} \qquad (6.50)$$

which can be re-written as:

$$p_T = \frac{1}{1 + (\tau_e - \tau_R) \cdot [-R(t)] + \tau_e} \qquad (6.51)$$

The objective is that of finding the optimal maintenance period τ^* which maximizes p_T. To this aim, we compute the derivative of p_T with respect to τ,

$$\frac{dp_T}{d\tau} = \frac{\tau_e \cdot R(\tau) + R^2(\tau) \cdot (\tau_R - \tau_e) - R'(\tau) \cdot (\tau_R - \tau_e) \cdot \int_0^\tau R(t)\,dt}{\left\{ \int_0^\tau R(t)\,dt + \tau_e \cdot [1 - R(\tau)] + \tau_R \cdot R(\tau) \right\}^2} \qquad (6.52)$$

Note that (Fig. 6.6):

$$R'(\tau) = \frac{dR(\tau)}{d\tau} \leq 0 \qquad \forall\, \tau > 0 \qquad (6.53)$$

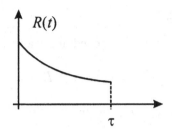

Fig. 6.6: Behavior of the reliability function $R(\tau)$

Then, we can distinguish two cases:

A. $\tau_R \geq \tau_e$

In this case, p_T is an increasing function of τ $\left(\dfrac{dp_T}{d\tau} > 0 \right)$, independently

of the functional form of the failure density $f(t)$, with or without aging.
Hence, there is no τ^* such that p_T achieves a maximum value.

B. $\tau_R < \tau_e$

In this case, it depends on the functional form of $f(t)$.

B.1   If $f(t)$ is such that $\dfrac{dp_T}{d\tau} > 0, \ \forall \tau > 0$, then no τ^* exists such that

p_T is maximum;

B.2   If $f(t)$ is such that $\dfrac{dp_T}{d\tau} = 0$, then

$$-\frac{R'(\tau^*)}{R(\tau^*)} \cdot \int_0^{\tau^*} R(t)\,dt - \left[1 - R(\tau^*)\right] = \frac{\tau_R}{\tau_e - \tau_R} \qquad (6.54)$$

from which the optimal value τ^*, can be determined.

Introducing the failure rate $\lambda(\tau^*)$:

$$\lambda(\tau^*) = \frac{\dfrac{dF}{dt}}{R} = -\frac{R'(\tau^*)}{R(\tau^*)} \tag{6.55}$$

we obtain the expression:

$$(\tau_e - \tau_R) \cdot \lambda(\tau^*) = \frac{\tau_R + (\tau_e - \tau_R) \cdot (1 - R(\tau^*))}{\displaystyle\int_0^{\tau^*} R(t)dt} \tag{6.56}$$

and the maximum value of p_T is:

$$p_T(\tau^*) = \frac{1}{1 + (\tau_e - \tau_R)\lambda(\tau^*)} \tag{6.57}$$

Example 6.2 [6]

Consider an exponential component, with failure time pdf $f(t) = \lambda \cdot e^{-\lambda t}$. Eq. (6.52) becomes:

$$\frac{dp_T}{d\tau} = \frac{\tau_R \cdot e^{-\lambda t}}{\left(\tau_e + \dfrac{1}{\lambda}\right) \cdot \left(1 - e^{-\lambda t}\right) + \tau_R \cdot e^{-\lambda t}} > 0$$

As expected, no τ^* exists for which p_T will achieve the maximum value. This means that no optimal preventive maintenance policy exists if the failure rate is constant: this is obvious since there is no aging.

Example 6.3 [6]

Consider a gamma component, with failure time pdf $f(t) = \lambda^2 \cdot t \cdot e^{-\lambda t}$. In this case the failure rate has the following expression:

$$\lambda(t) = \frac{\lambda^2 \cdot t}{1 + \lambda \cdot t}$$

and the reliability function:

$$R(t) = (1 + \lambda \cdot t) \cdot e^{-\lambda t}$$

Maximizing the average availability p_T:

$$\frac{dp_T}{d\tau} = 0$$

we obtain the optimal value τ^* in correspondence of which the average availability p_T is:

$$p_T(\tau^*) = \frac{1 + \lambda \cdot \tau^*}{1 + \lambda \cdot \tau^* + (\tau_e - \tau_R) \cdot \lambda^2 \cdot \tau^*}$$

6.9 A policy of preventive replacement with economical optimization

Before proceeding with the development of a replacement model, it is important to note that preventive replacement actions, that is, those taken before the equipment reaches a failed state, require two necessary conditions:

a. The total cost of the replacement must be greater after failure than before (if "cost" is the appropriate criterion – otherwise the appropriate criterion, such as UPtime, is substituted in place of cost). This may be caused by a greater loss of production since replacement after failure is unplanned or failure of one piece of plant may cause damage to other equipment.

b. The failure rate $\lambda(t)$ of the equipment must be increasing. Note, however, that preventive maintenance of a general nature which does not return equipment to the as new condition, may be appropriate for equipment subject to a constant failure rate.

Determination of the best level of such preventive work is related to the problem of determination of the optimal frequency of inspection and minor maintenance of complex equipment.

We will now deal with the calculation of the optimal preventive replacement of equipment subject to breakdown [6]. The time at which the preventive replacement occurs, depends on the age of the equipment. When failures occur, failure replacements are made. The problem is to balance the cost of preventive replacement against their benefits and we do this by determining the optimal preventive replacement age t_p^* for the equipment to minimize the total expected cost of replacements per unit time. Let C_p be the cost of preventive replacement, C_f the cost of a replacement at failure and $f(t)$ the probability density function of the failure times of the equipment. The replacement policy is to perform a preventive replacement once the equipment has reached a specified age t_p and failure replacements when necessary. This policy is illustrated in Fig. 6.7.

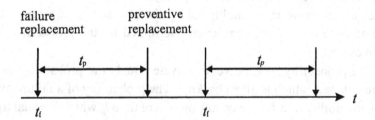

Fig. 6.7: Preventive replacement policy [6]

The objective is to determine the optimal replacement age t_p^* of the equipment which minimizes the total expected replacement cost per unit time.

In this problem, there are two possible cycles of operation: one cycle being determined by the equipment reaching its planned replacement age t_p, the other being determined by the equipment ceasing to operate due to a failure occurring before the planned replacement time. This two possible cycles are illustrated in Fig. 6.8.

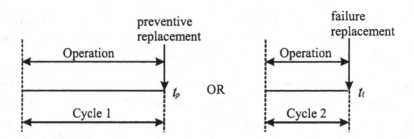

Fig. 6.8: Possible cycles under the preventive replacement policy [6]

Based on these two possible cycles, the total expected cost of replacement per unit time $C(t_p)$ is computed as the fraction between the total expected replacement cost per cycle and the expected cycle length. The total expected replacement cost per unit time $C(t_p)$ is given by:

$$C(t_p) = \frac{Total\ expected\ replacement\ cost\ per\ cycle}{Expected\ cycle\ length} \qquad (6.58)$$

The total expected replacement cost per cycle is equal to the sum of the cost of a preventive cycle multiplied by the probability of a preventive cycle and the cost of a failure cycle multiplied by the probability of a failure cycle.

The probability of a preventive cycle equals the probability of no failure before t_p which is given by $R(t_p)$. The probability of a failure cycle is the probability of a failure occurring before time t_p which is equal to:

$$F(t_p) = 1 - R(t_p) \qquad (6.59)$$

The expected cycle length is equal to the sum of the length of a preventive cycle multiplied by the probability of a preventive cycle plus the expected length of a failure cycle multiplied by the probability of a failure cycle, i.e.

$$t_p \cdot R(t_p) + M(t_p) \cdot [1 - R(t_p)] \qquad (6.60)$$

where $M(t_p)$ denotes the length of a failure cycle.

As t_p is the maximum failure time, the distribution of the failure times has to be conditioned by the probability of a failure cycle which is $1 - R(t_p)$. So the expected length of a failure cycle is given by:

$$M(t_p) = \int_0^{t_p} t \cdot \frac{f(t)dt}{1 - R(t_p)}$$
(6.61)

Summarizing, the total expected replacement cost per unit time $C(t_p)$ is given by:

$$C(t_p) = \frac{C_p \cdot R(t_p) + C_f \cdot (1 - R(t_p))}{t_p \cdot R(t_p) + M(t_p) \cdot (1 - R(t_p))}$$
(6.62)

This model relates the replacement age t_p to the total expected replacement cost per unit time. A variation to this model is for example taking into account the time required to perform a failure or a preventive replacement.

References

[1] Rausand, M. and Hoyland, A., System Reliability Theory, Wiley, 2004.
[2] Ushakov, I.A., Handbook of Reliability Engineering, Wiley, 1994.
[3] Schneeweiss, W.G., Reliability Modeling, LiLoLe-Verlag, 2001.
[4] Birolini, A., Reliability Engineering, Springer, 2004.
[5] Lewis, E.E., Introduction to Reliability Engineering, Wiley, 1996.
[6] Badoux, R.A.J., Availability and Maintainability, In Reliability Modeling and Applications, A.G. Colombo and A.Z. Keller Eds., Kluwer, 1986.

7

Fault Tree Analysis

7.1 Introduction

For complex multi-component systems, for example such as those employed in the nuclear, chemical, process and aerospace industries, it is important to analyze the possible mechanisms of failure and to perform probabilistic analyses for the expected frequency of such failures. Often, each such system is unique in the sense that there are no other identical systems (same components interconnected in the same way and operating under the same conditions) for which failure data have been collected: therefore a statistical failure analysis is not possible. Furthermore, it is not only the probabilistic aspects of failure of the system which are of interest but also the initiating causes and the combination of events which can lead to a particular failure.

The engineering way to tackle a problem of this nature, where many events interact to produce other events, is to relate these events using simple logical relationships (intersection, union, etc.) and to methodically build a logical structure which represents the system.

In this respect, *Fault tree analysis* is a systematic, deductive technique which allows to develop the causal relations leading to a given undesired event. It is deductive in the sense that it starts from a defined system failure event and unfolds backward its causes down to the primary (basic) independent faults. The method focuses on a single system failure mode and can provide qualitative information on how a particular event can occur and what consequences it leads to, while at the same time allowing the identification of those components which play a major role in determining the defined system failure. Moreover it can be solved in quantitative terms to provide the probability of events of interest starting from knowledge of the probability of occurrence of the basic events which cause them.

In the following, we shall give only the basic principles of the technique. The interested reader is invited to look at the specialized literature for further details, e.g. [1], [2], [3], [4], [5], [6], [6], [8], [9], [10], [11], [12], [13] from which most of the material herein contained has been taken.

7.2   Fault tree construction

A fault tree is a graphical representation of causal relations obtained when a system failure mode is traced backward to search for its possible causes. To complete the construction of a fault tree for a compex system, it is necessary to first understand how the system functions. A system flow diagram (e.g. a reliability block diagram) is used for this purpose, e.g. to depict the pathways by which materials are transmitted between components of the system.

The first step in fault tree construction is the selection of the system failure event of interest. This is called the *top event* and every following event will be considered in relation to its effect upon it.

The next step is to identify contributing events that may directly cause the top event to occur. At least four possibilities exist [5]:

1. no input to the device;
2. primary failure of the device (under operation in the design envelope, random, due to aging or fatigue);
3. human error in actuating or installing the device;
4. secondary failure of the device (due to present or past stresses caused by neighboring components or the environments: e.g. common cause failure, excessive flow, external causes such as earthquakes).

If these events are considered to be indeed contributing to the system fault, then they are connected to the top event logically via an OR function and graphically through the OR gate (Fig. 7.1):

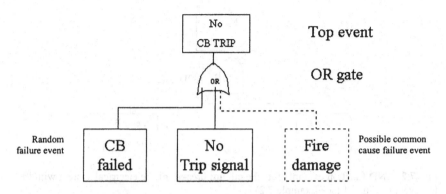

Fig. 7.1: Top and first level of a fault tree for a circuit breaker (CB) failing to trip an electrical circuit (see Example 7.2) [14]

Once the first level of events directly contributing to the top has been established, each event must be examined to decide whether it is to be further decomposed in more elementary events contributing to its occurrence. At this stage, the questions to be answered are:

1. is this event a primary failure?
2. is it to be broken down further in more primary failure causes?

In the first case, the corresponding branch of the tree is terminated and this primary event is symbolically represented by a circle. This also implies that the event is independent of the other terminating events (circles) which will be eventually identified and that a numerical value for the probability of its occurrence is available if a quantitative analysis of the tree is to be performed.

On the contrary, if a first level contributing event is not identified as a primary failure, it must be examined to identify the sub-events which contribute to its occurrence and their logical relationships (Fig. 7.2).

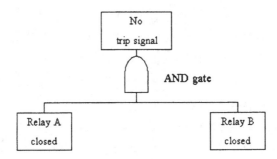

Fig. 7.2: AND function example for the circuiti breaker of the electrical system with the top event of Fig. 7.1 (see Example 7.2) [14]

The procedure of analyzing every event is continued until all branches have been terminated in independent primary failures for which probability data are available. Sometimes, certain events which would require further breakdown can be temporarily classified as primary at the current state of the tree structure and assigned a probability by rule of thumb. These underdeveloped events are graphically represented by a diamond symbol rather than by a circle (see Example 7.1 below).

Example 7.1: mechanical holding latch

Consider the failure of the mechanical holding latch of Figure 7.3(a). The corresponding fault tree is given in Figure 7.3(b).

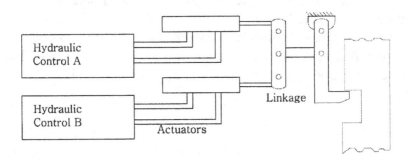

Fig. 7.3(a): Mechanical holding latch [5]

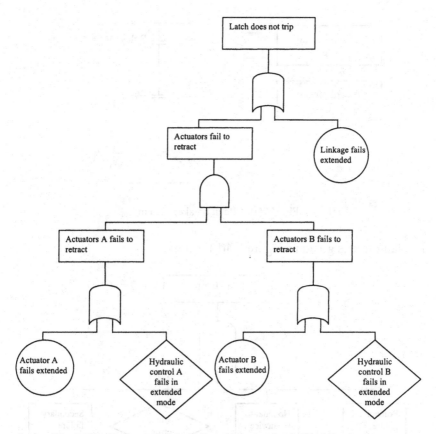

Fig. 7.4(b): Fault tree for the failure of the mechanical holding latch

Example 7.2: Circuit breaker trip

Draw the fault tree for the failure to trip of the circuit breaker shown in Figure 7.4(a). The circuit breaker opens when there is no voltage across the UV trip coil. Assume for simplicity that of all components only the UV trip coil cannot fail.

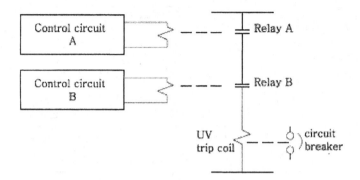

Fig. 7.5(a): Circuit breaker system

The fault tree is given in Figure 7.4(b) below.

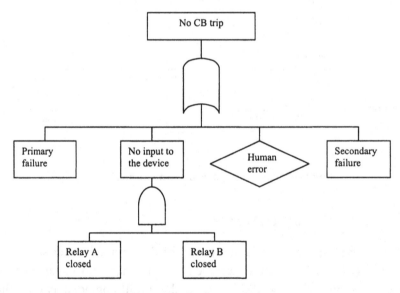

Fig. 7.4(b): Fault tree solution

Tables 7.1 and 7.2 report the symbols employed to represent the events and their relationships in a fault tree.

Table 7.1: Event Symbols

Event Symbol	Meaning of Symbol
⭘	Basic event with sufficient data
◇	Undeveloped event
▢	Event represented by a gate
⬯	Condition event used with inhibit gate
⌂	House event. Either occurring or not occurring
◁ △	Transfer symbol

Table 7.2: Gate Symbols

Gate Symbol	Gate Name	Causal Relation
	AND gate	Output event occurs if all input events occur simultaneously.
	OR gate	Output event occurs if any one of the input events occurs.
	Inhibit gate	Input produces output when conditional event occurs.
	Priority AND Gate	Output event occurs if all input events occur in the order from left to right.
	Exclusive OR Gate	Output event occurs if one, but not both, of the input events occur.
 n input	m out of n gate (volume or sample gate)	Output event occurs if m out of n input events occur.

It is interesting to note that all the more complicated gate symbols can be constructed with the basic AND, OR and NOT symbols. Some examples are presented in Figs. 7.6-7.12 [5].

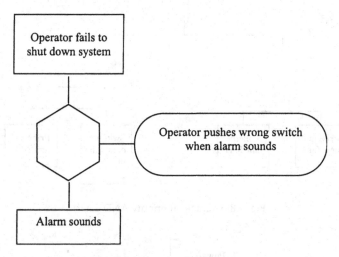

Fig. 7.6: Example of inhibit gate

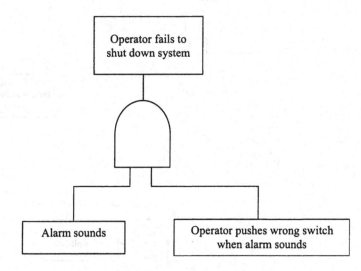

Fig. 7.7: Equivalent logical form to Fig. 7.6

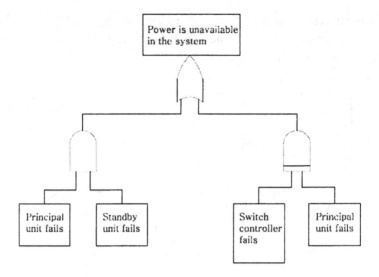

Fig. 7.8: Example of priority AND gate

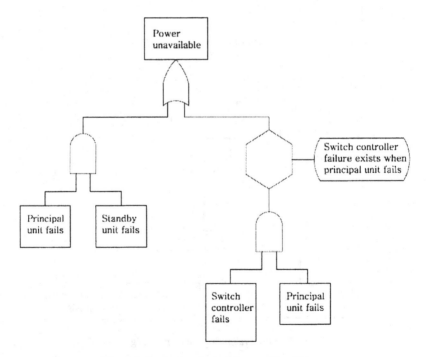

Fig. 7.9: Equivalent logical form to Fig. 7.8

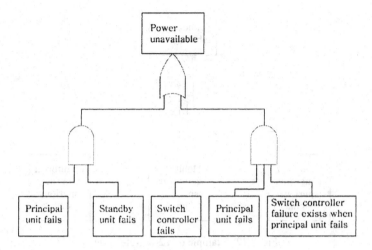

Fig. 7.10: Equivalent logical form to Fig. 7.8

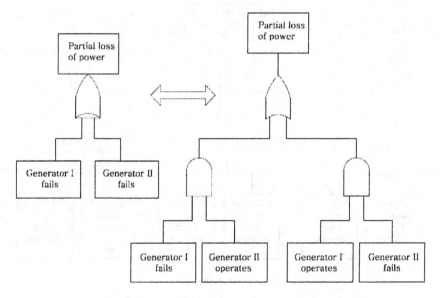

Fig. 7.11: Example of exclusive OR gate and its equivalent logical expression

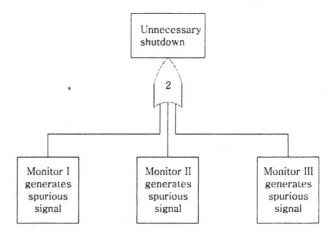

Fig. 7.12: Example of a 2-out-of-3 gate

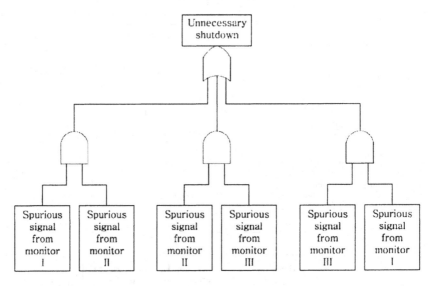

Fig. 7.13: Equivalent logical form to Fig 7.11

Actual construction of fault trees is an art as well as a science and comes mainly through experience. Below some useful guidelines are reported [6].

Rule 1. State the fault event as a fault, including the description and timing of a fault condition at some particular time. Include:
(a) what the fault state of that system or component is;
(b) when that system or component is in the fault state.
Test the fault event by asking:
(a) Is it a fault?
(b) Is the what-and-when portion included in the fault statement?

Rule 2. There are two basic types of fault statements, state-of-system and state-of-component. To continue the tree:
(a) if the fault statement is a state-of-system statement, use Rule 3;
(b) if the fault statement is a state-of-component statement, use Rule 4.

Rule 3. A state-of-system fault may use an AND, OR, or INHIBIT gate or no gate at all. To determine which gate to use, the faults must be the:
(a) minimum necessary and sufficient fault events;
(b) immediate fault events. To continue, state the fault events input into the appropriate gate.

Rule 4. A state-of-component fault always uses an OR gate. To continue, look for the primary, secondary, and command failure fault events. Then state those fault events:
(a) primary failure is failure of that component within the design envelope or environment;
(b) secondary failures are failures of that component due to excessive environments exceeding the design environment;
(c) command faults are inadvertent operation of the component because of a failure of a control element.

Rule 5. No gate-to-gate relationships, i.e., put an event statement between any two gates.

Rule 6. Expect no miracles; those things that would normally occur as the result of a fault will occur, and only those things. Also, normal system operation may be expected to occur when faults occur.

Rule 7. In an OR gate, the input does not cause output. If any input exists, the output exists. Fault events under the gate may be a restatement of the output events.

Rule 8. An AND gate defines a causal relationship. If the input events coexist, the output is produced.

Rule 9. An INHIBIT gate describes a causal relationship between one fault and another, but the indicated condition must be present. The fault is the direct and sole cause of the output when that specified condition is present. Inhibit conditions may be faults or situations, which is why AND and INHIBIT gates differ.

7.3   Qualitative analysis: coherent structure functions and minimal cut sets

7.3.1   Structure functions

A fault tree can be described by a set of Boolean algebraic equations, one for each gate of the tree. For each gate, the input events are the independent variables and the output event is the dependent variable. Utilizing the rules of Boolean algebra it is then possible to solve these equations so that the top event is expressed in terms of sets of primary events only.

When dealing with a Boolean event E_j, we can introduce an indicator variable X_j which is equal to 1 if the event is true and 0 if it is false. If the system and components are considered from the point of view of reliability then $X_j = 1$ indicates success and $X_j = 0$ failure; viceversa from the point of view of safety.

The top event of a fault tree can be represented by an indicator variable X_T which is a Boolean function of the Boolean variables $X_1, X_2, ..., X_n$ describing the states of the n events of the system:

$$X_T = \Phi(X_1, X_2, ..., X_n) \tag{7.1}$$

Such function is called a *switching* or *structure* function and incorporates all the causal relations among the events which lead to the top event. It maps an n-dimensional vector $\underline{X} = (X_1, X_2, ..., X_n)$ of 0's and 1's onto a binary variable equal to 0 or 1. For example, looking at a simple series system from the reliability viewpoint, we have that its success occurs when all its components are in a success state. From the rules of Boolean algebra, the corresponding structure function is:

$$X_T = \prod_{j=1}^{n} X_j \tag{7.2}$$

For a parallel system, at least one of the components must be in the success state for the system to be successful. Correspondingly, we have:

$$X_T = 1 - (1 - X_1)(1 - X_2)..(1 - X_n) = \coprod_{j=1}^{n} X_j \tag{7.3}$$

Obviously, for a given system there are various forms which can be used to write the structure function. The task that we wish to undergo is that of using the rules of Boolean algebra to reduce a structure function to its most simplified equivalent version.

First of all, we introduce the concept of *fundamental product* which is a product containing all of the n input variables, complemented or not. For n variables there are 2^n such products; for example, for $n=3$, we have:

$$X_1 X_2 X_3, X_1 X_2 \overline{X}_3, X_1 \overline{X}_2 X_3, \overline{X}_1 X_2 X_3, \overline{X}_1 X_2 \overline{X}_3, \overline{X}_1 \overline{X}_2 X_3, X_1 \overline{X}_2 \overline{X}_3, \overline{X}_1 \overline{X}_2 \overline{X}_3 \tag{7.4}$$

Clearly a fundamental product is 1 if and only if all its variables are 1.

An important theorem states that a structure function can be written uniquely as the union of the fundamental products which correspond to the combinations of the variables which render the function true (i.e., $\Phi = 1$). This is called the canonical expansion or disjunctive normal form of Φ [2].

Using the rules of Boolean algebra (see Table 7.3), the canonical expansion can be simplified further to obtain an irreducible expression of the structure function in terms of *minimal cutsets*.

Table 7.3: Some rules of Boolean algebra for events

1)   Commutative Law:
- (a) $XY = YX$
- (b) $X + Y = Y + X$

2)   Associative Law
- (a) $X(YZ) = (XY)Z$
- (b) $X + (Y + Z) = (X + Y) + Z$

3)   Idempotent Law
- (a) $XX = X$
- (b) $X + X = X$

4)   Absorption Law
- (a) $X(X + Y) = X$
- (b) $X + XY = X$

5)   Distributive Law
- (a) $X(Y + Z) = XY + XZ$
- (b) $(X + Y)(X + Z) = X + YZ$

6)   Complementation*
- (a) $X\overline{X} = \varnothing$
- (b) $X + \overline{X} = \Omega$
- (c) $\overline{\overline{X}} = X$

7)   Unnamed relationships but frequently useful
- (a) $X + \overline{X}Y = X + Y$
- (b) $\overline{X}(X + Y) = \overline{X}Y$

*The universal event Ω iss sometimes denoted by I, and the null event $\varnothing$ is sometimes denoted by 0.

7.3.2   Coherent structure functions and minimal cut sets

A physical system would be quite unusual (or perhaps poorly designed) if improving the performance of a component (that is, replacing a failed component by a functioning one) caused the system to change from the success to the failed state. Thus, we restrict consideration to structure functions that are monotonically increasing in each input variable. These structure functions do not contain complemented variables; they are called *coherent* and can always be expressed as the union of fundamental products.

The main properties of a coherent structure function are:

1. $\Phi(\underline{1}) = 1$ if all the components are in their success state, the system is successful;
2. $\Phi(\underline{0}) = 0$ if all the components are failed, the system is failed;
3. $\Phi(\underline{X}) \geq \Phi(\underline{Y})$ for $\underline{X} \geq \underline{Y}$

The last property accounts for the fact that considering two distinc system configurations, represented by the indicator variable $\underline{X}$ and $\underline{Y}$ if $\Phi(\underline{Y}) = 1$ and a failed component in $\underline{Y}$ is repaired in $\underline{X}$, this cannot cause the system to fail ($\Phi(\underline{X}) = 1$); in other words if the system in $\underline{Y}$ was failed ($\Phi(\underline{Y}) = 0$), in $\underline{X}$ it can either remain failed or be repaired ($\Phi(\underline{X}) = 1$); otherwise, if the system in $\underline{Y}$ was successful ($\Phi(\underline{Y}) = 1$), the additional repair can only make it maintain its successful status.

Coherent structure functions can be expressed in reduced expressions in terms of minimal path or cut sets. A path set is a set $\underline{X}$ such that $\Phi(\underline{X}) = 1$; a cut set is a set $\underline{X}$ such that $\Phi(\underline{X}) = 0$. Physically, a path (cut) set is a set of components whose functioning (failure) ensures the functioning (failure) of the system.

A minimal path (cut) set is a path (cut) set that does not have another path (cut) set as a subset. Physically, a minimal path (cut) set is an irreducible path (cut) set: failing (repairing) one element of the set fails (repairs) the system. Therefore, removing one element from a path (cut) set makes the set thereby obtained no longer a path (cut) set. Once the path (cut) sets are identified, the system structure function can be expressed as the union of the path (cut) sets: this constitutes a unique and irreducible form of the coherent structure function of the system.

From this analysis we see that any fault tree can be equivalently written in a form with an OR gate in the first level below the top event combining all the minimal cut sets, each one in turn represented by an AND gate intersecting all the elements comprising the given minimal cut set.

For trees of systems with relatively few components, the minimal cut sets can be identified by inspection. Most often, however, such an approach is very inefficient, if possible at all, since the number of minimal cut sets increases very rapidly as the complexity of the tree increases. Therefore, a more systematic approach should be undertaken by which after writing the Boolean equations for each gate, Boolean algebra is used to solve the top event structure function in terms of the cut sets; using again Boolean algebra one can then eliminate all the redundancies in the events to obtain the minimal cut sets. Several computerized approaches exist to perform this task.

After the minimal cut sets have been obtained, the qualitative analysis is complete and the failure modes contributing to the top event have been identified. The analysis provides us with some indications on the criticality of the various components: those appearing in minimal cut sets of low order (number of primary events constituting the cut set) and those most frequently appearing in the various cut sets are good candidate to be critical for the system safe operation.

Two general rules of thumb for judging the importance of a minimal cut set are :

1. the importance of a minimal cut set is inversely proportional to its order;
2. any one-event minimal cut set should be avoided by re-design if possible.

7.4   Quantitative analysis

Quantitative analysis of the fault tree consists of transforming its logical structure into an equivalent probability form and numerically calculating the probability of occurrence of the top event from the probabilities of occurrence of the basic events. The probability of the basic event is the failure probability of the component or subsystem during the mission time of interest.

From the definition of the structure function $\Phi(\underline{X})$ as a function of the indicator variables of the basic events $\underline{X} = (X_1, X_2, ..., X_n)$, we see that the structure function is itself an indicator variable which is equal to 1 when the top event is verified and 0 otherwise. Consequently we may write, for the probability of the top event:

$$P(\Phi(\underline{X}) = 1) = E[\Phi] = 0 \cdot P(\Phi(\underline{X}) = 0) + 1 \cdot P(\Phi(\underline{X}) = 1) \qquad (7.5)$$

where $E[\cdot]$ is the expectation operator. Given the expression of the structure function Φ in terms of the indicator variables of the basic events, it is possible to write the probability (7.5) in terms of the probability values of the independent basic events, $P(X_i = 1) = E[X_i]$.

Consistent with what previously said concerning the qualitative analysis of fault trees, there exist two approaches for calculating the probability of the top event from the probabilities of the basic events. If the fault tree is not solved for the minimal cut sets, then the probability of the top event can be calculated by hand, provided that the size and complexity of the tree are not too large. This is done proceeding in an orderly fashion from the bottom to the top of the tree and computing at each gate the probability of the output from the probabilities of the input events, using the laws of probability corresponding to that gate structure (AND, OR, etc.). This can be "automatically" done through Eq. (7.5). For example, the probability of the output Y of an AND gate with two independent input events X_1, X_2, with probability P_1 and P_2 respectively, is

$$P(Y = 1) = E[X_1 X_2] = E[X_1] \cdot E[X_2] = P_1 P_2 \qquad (7.6)$$

while for the output of an OR gate,

$$P(Y = 1) = E\left[\left(1 - (1 - X_1)(1 - X_2)\right)\right] =$$
$$= E[X_1 + X_2 + X_1 X_2] = E[X_1] + E[X_2] + E[X_1 X_2] = P_1 + P_2 - P_{12} \tag{7.7}$$

where P_{12} is the probability of the intersection event $X_1 X_2 = 1$, given by (7.6) in the case that X_1 and X_2 are independent.

On the contrary, if a qualitative analysis has been performed to determine the system minimal cut sets $M_1, M_2, ..., M_{mcs}$, by definition the probability of each of them is the probability of the intersection of the independent basic events comprising that minimal cut set, i.e.,

$$P(M_i) = P(X_1^i)P(X_2^i)... \qquad i = 1, 2, ..., mcs \tag{7.8}$$

where the product is extended to all the events comprising M_i. By definition, the system structure function is the intersection of the *mcs* minimal cut sets:

$$\Phi(\underline{X}) = 1 - (1 - M_1)(1 - M_2)..(1 - M_{mcs}) = \coprod_{j=1}^{mcs} M_j \tag{7.9}$$

and the probability of the top event is

$$P(\Phi(\underline{X}) = 1) = E\left[\sum_{j=1}^{mcs} M_j - \sum_{i=1}^{mcs-1}\sum_{j=i+1}^{mcs} M_i M_j + ... + (-1)^{mcs+1} \prod_{j=1}^{mcs} M_j\right] =$$
$$= E\left[\sum_{j=1}^{mcs} M_j\right] - E\left[\sum_{i=1}^{mcs-1}\sum_{j=i+1}^{mcs} M_i M_j\right] + ... + (-1)^{mcs+1} E\left[\prod_{j=1}^{mcs} M_j\right] =$$
$$= \sum_{j=1}^{mcs} P(M_j) - \sum_{i=1}^{mcs-1}\sum_{j=i+1}^{mcs} P(M_i M_j) + ... + (-1)^{mcs+1} P\left(\prod_{j=1}^{mcs} M_j\right) \tag{7.10}$$

For two minimal cut sets, the formula gives the well-known result (7.7),

$$P(\Phi(\underline{X}) = 1) = P(M_1) + P(M_2) - P(M_1 M_2) \tag{7.11}$$

It can be shown that the following upper and lower bounds to eq. (7.10) hold (Eqs. (4.11) and (4.12) in Section 4.4.1):

$$P(\Phi(\underline{X}) = 1) \le \sum_{j=1}^{mcs} P(M_j)$$

$$P(\Phi(\underline{X}) = 1) \ge \sum_{j=1}^{mcs} P(M_j) - \sum_{i=1}^{mcs-1} \sum_{j=i+1}^{mcs} P(M_i M_j)$$

(7.12)

In reliability and risk calculations, basic events are typically rare (low probability events), so that the probability of their intersection in minimal cut sets , i.e. that some of them are verified simultaneously so as to verify a minimal cut set, is very small; therefore, one can approximate using the first of the eq. (7.12) (*rare-event approximation,* Section 4.4.1):

$$P(\Phi(\underline{X}) = 1) \cong \sum_{j=1}^{mcs} P(M_j)$$

(7.13)

7.5 Comments

Although actual construction of fault trees is an art as well as a science and comes only through experience, fault tree analysis is a widely adopted tool for safety and risk analyses. Some of its recognized advantages are:

1. Straightforward modelization via few, simple logic operators;
2. Directing the analysis to ferret out failures;
3. Focus on one top event of interest at a time;
4. Pointing out the aspects of the system important to the failure of interest;
5. Providing a graphical communication tool whose analysis is transparent;
6. Providing an insight into system behaviour;
7. Minimal cut sets are a synthetic result which identifies the critical components.

References

[1] Aven, T., Foundations of Risk Analysis, Wiley, 2003.

[2] Barlow, R., Engineering Reliability, ASA-SIAM series on Statistics and Applied Probability, 1998.

[3] Bedford, T. and Cooke, R., Probabilistic Risk Analysis, Cambridge University Press, 2001.

[4] Fault Tree Handbook with Aerospace Applications, NASA, 2002.

[5] Henley, E.J. and Kumamoto, H., Probabilistic risk assessment, NY, IEEE Press, 1992.

[6] Lambert, H. E., Systems Safety Analysis and Fault Tree Analysis, Lawrence Livermore Laboratory Rep. UCID-16238, 1973.

[7] McCormick, N.J., Reliability and risk analysis, New York, Academic Press, 1981.

[8] PRA Procedures Guide, Vols. 1&2, NUREG/CR-2300, 1983.

[9] Probabilistic Risk Assessment Procedures Guide for NASA Managers and Practitioners, NASA, 2002.

[10] Rao, S.S., Reliability-based design, NY , McGraw-Hill, 1992.

[11] Rausand, M. and Hoyland, A., System Reliability Theory, Wiley, 2004.

[12] WASH-1400, Reactor Safety Study, 1975.

[13] Schneeweiss, W.G., The Fault Tree Method, LiLoLe, 1999.

[14] Reliability Manual for Liquid Metal Fast Reactor (LMFBR) Safety Programs, General Electric Company Internal Rep. SRD-74-113, 1974.

8

Event Tree Analysis

8.1 Introduction

Event trees are inductive logic methods for identifying the various accident sequences which can generate from a single initiating event. The approach is based on the discretization of the real accident evolution in few macroscopic events. The accident sequences which derive are then quantified in terms of their probability of occurrence.

The events delineating the accident sequences are usually characterized in terms of: i) the intervention (or not) of protection systems which are supposed to take action for the mitigation of the accident (*system event tree*); ii) the fulfillment (or not) of safety functions (*functional event tree*); iii) the occurrence or not of physical phenomena (*phenomenological event tree*).

Typically, the functional event trees are an intermediate step to the construction of system event trees: following the accident-initiating event, the safety functions which need to be fulfilled are identified; these will later be substituted by the corresponding safety and protection systems.

The system event trees are used to identify the accident sequences developing within the plant and involving the protection and safety systems.

The phenomenological event trees describe the accident phenomenological evolution outside the plant (fire, contaminant dispersion, etc.).

In the following, we shall give only the basic principles of the technique. The interested reader is invited to look at the specialized literature for further details, e.g. [1], [2], [3], [4], [5], [6], [7], [8], from which most of the material herein contained has been taken.

8.2   Event tree construction

An event tree begins with a defined accident-initiating event which could be a component or an external failure. It follows that there is one event tree for each different accident-initiating event considered. This aspect obviously poses a limitation on the number of initiating events which can be analyzed in details. For this reason, the analyst groups similar initiating events and only one representative initiating event for each class is investigated in details. Initiating events which are grouped in the same class are usually such to require the intervention of the same safety functions and to lead to similar accident evolutions and consequences.

Once an initiating event is defined, all the safety functions that are required to mitigate the accident must be defined and organized according to their time of intervention. For example (Fig. 8.1) if the initiating event (I) is the rupture of a tube with release of inflammable liquid and the sparking of jet-fire, the first function required would be that of interception of the released flow rate, followed by the cooling of adjacent tanks and finally the quenching of the jet. These functions are structured in the form of headings in the functional event tree. For each function, the set of possible success and failure states must be defined and enumerated. Each state gives rise to a branching of the tree (Fig. 8.1). For example, in the typical binary success/failure logic it is customary to associate to the top branch the success of the function and to the bottom branch its failure.

Besides the time-order, also the logic order of the required functions must be accounted for. In other words, if the successful fulfillment of a given function is dependent on the fulfillment of another one, the tree needs to be re-ordered in such a way that the dependent functions follow those upon which they depend. This allows pruning of some sequences. Consider, for example, a dependent function S_1 whose fulfillment depends on the success of a function S_2; then, the branch following the failure of S_2 needs not be further decomposed in two branches for S_1 successful or not, because failure of S_2 implies no fulfillment of S_1 (Fig. 8.2).

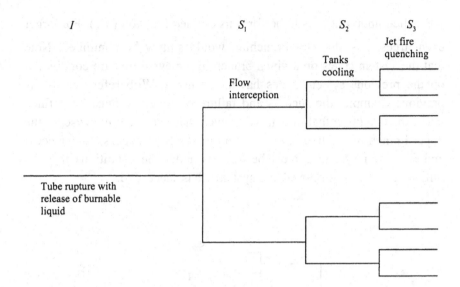

Fig. 8.1: Example of functional event tree

The functions in the tree are then substituted by the safety systems which must perform them: again respecting the logical dependencies may lead to additional pruning. System dependencies can be functional, if the failure of intervention of a system renders helpless the intervention of the successive one, or structural if the systems share some common parts or flow so that malfunctioning of that part makes them both fail.

Once the system failure and success states have been properly defined, the states are combined through the tree branching logic to obtain the various accident sequences that are associated with the given initiating event.

Fig. 8.3 shows a graphical example of a system event tree: the initiating event is depicted by the initial horizontal line and the system states are then connected in a stepwise, branching fashion: system success and failure states have been denoted by S and F, respectively. The accident sequences that result from the tree structure are shown in the last column. Each branch yields one particular accident sequence; for example, IS_1F_2 denotes the accident sequence in which the initiating event (I) occurs, system 1 is called upon and succeeds (S_1), and system

2 is called upon but fails to perform its defined function (F_2). For larger event trees, this stepwise branching would simply be continued. Note that the system states on a given branch of the event tree are conditional on the previous system states having occurred. With reference to the previous example, the success and failure of system 1 must be defined under the condition that the initiating event has occurred; likewise, in the upper branch of the tree corresponding to system 1 success, the success and failure of system 2 must be defined under the conditions that the initiating event has occurred and system 1 has succeeded.

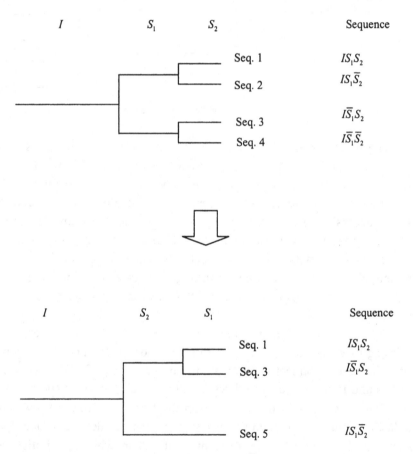

Fig. 8.2: Functional dependences: the negated events $\overline{S}_i$, i=1,2, denote failure of the corresponding function

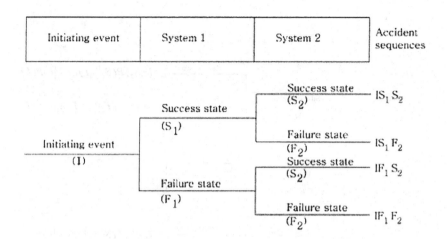

Fig. 8.3: Illustration of event tree branching [8]

8.3 Event tree evaluation

Once the final event tree has been constructed, the final task is to compute the probabilities of system failure. Each event (branch) in the tree can be interpreted as the top event of a fault tree which allows the evaluation of the probability of the occurrence of such event. The value thus computed represents the conditional probability of the occurrence of the event, given that the events which precede on that sequence have occurred. Multiplication of the conditional probabilities for each branch in a sequence gives the probability of that sequence (Fig. 8.4).

In the case of structural dependencies, two approaches to accident sequence modelling are available [5]. One approach is called *event tree with boundary conditions* and consists in decomposing the system so as to identify the supporting parts or functions upon which some components and systems are simultaneously dependent. The supporting parts thereby identified appear explicitly as system event tree headings, preceding the dependent protection systems and components. Since dependent parts are extracted and explicitly treated as boundary conditions in the event tree, this approach leads to large fault trees and relatively small event trees.

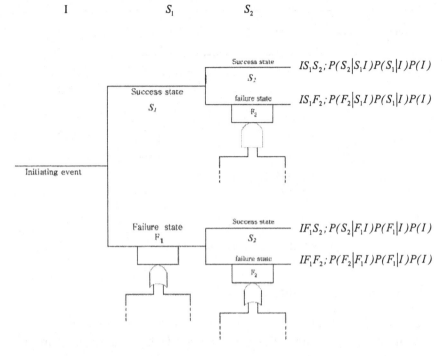

Fig. 8.4: Schematics of the event tree shown with the fault trees used to evaluate the probabilities of different events

For example, consider an initiating event which requires two systems, S_1 and S_2, to intervene and suppose that S_1 needs the pumps of S_2 to operate. Then, one could extract the common part and consider three systems: S_1, S_2^*, which is the S_2 system without the pumps common to S_1, and S_3, which represents the pumps used by both S_1 and S_2 (Fig. 8.5). Then, the dependencies are explicitly represented in the tree and the branching associated to $S_1^{'}$ and S_2^* is eliminated when S_3 is not functioning. Thus, all the conditional probabilities are independent and the probability of the accident sequences can be computed by simple multiplication. This way of proceeding simplifies considerably the computations but it requires a great deal of expertise by the analyst. In fact, since system interactions and dependencies are treated primarily within the inductive logic of the event tree, those dependencies not recognized by the analyst may not be incorporated into the analysis.

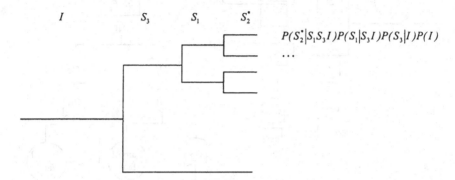

$$P(S_2^*|S_1S_3I)P(S_1|S_3I)P(S_3|I)P(I)$$
...

Fig. 8.5: Event tree with boundary conditions

The second approach is called *Fault-tree link* [5]. In this method, the dependencies from support systems or common parts are modeled in the fault trees, so that at the level of the event tree the systems are inserted without any care of their structural dependencies. For each sequence of the event tree, then, the fault trees of the composing events are linked in one large fault tree which follows the logic depicted in the event tree and the large fault tree is then solved with the usual techniques to compute the probability of occurrence of that sequence.

Fig. 8.6 shows the previous example of Fig. 8.5. Only systems S_1 and S_2 are explicited on the event tree without particular care to their dependence. If we now want to evaluate the probability of the sequence IS_1S_2, we build a fault tree whose top event occurs when the initiating event I and the failure of both systems S_1 and S_2 occur. In place of the events S_1 and S_2 we can substitute their corresponding system fault trees, thus obtaining a large fault tree which can be logically simplified (accounting for the existing dependencies) and evaluated so as to give the probability of the top event, i.e. the probability of the sequence of interest. With this method, the dependencies are properly treated even if the analyst was, a priori, unaware that they existed. On the other hand, the resulting fault tree for an accident sequence may be rather large.

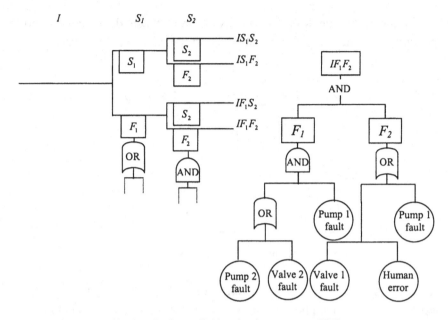

Fig. 8.6: Fault tree linking for the sequence IF_1F_2

In summary, in the event trees with boundary conditions all the significant dependencies among systems are explicitly represented in the event tree; the fault trees for the individual events are then simple and independent but the analyst must take great care in identifying all the existing dependencies. In the fault tree-link approach, dependencies are included in the fault trees for the various systems and thus they are not dependent; the linked fault tree of a generic accident sequence of interest is rather large and complex but all dependencies are treated automatically.

Finally, in Fig. 8.7 we report a simplified version of a functional event tree for the case of a large break of a pipe in the primary cooling circuit of a nuclear reactor.

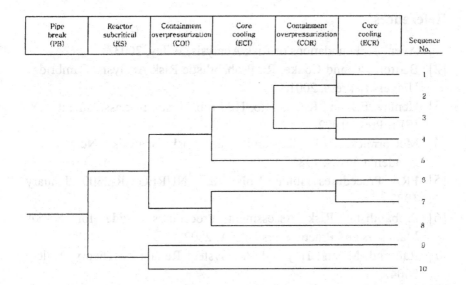

| | Pipe break (PB) | Reactor subcritical (RS) | Containment overpressurization (COI) | Core cooling (ECI) | Containment overpressurization (COR) | Core cooling (ECR) | Sequence No. |

Seq. No.	RS	COI	ECl	COR	ECR	Remarks
1						Core cooled
2					f	Slow melt
3						Core cooled
4				f	f	Slow melt
5			f	NA	NA	Melt
6		f		NA		Core cooled
7		f		NA	f	Slow melt
8		f	f	NA	NA	Melt
9	f		NA	NA	NA	Melt
10	f	f	NA	NA	NA	Melt

f = function failure; NA = not applicable.

Fig. 8.7: Functional event tree for a large break LOCA (Loss of Coolant Accident) in a nuclear reactor

References

[1] Aven, T., Foundations of Risk Analysis, Wiley, 2003.

[2] Bedford, T. and Cooke, R., Probabilistic Risk Analysis, Cambridge University Press, 2001.

[3] Henley, E.J. and Kumamoto, H., Probabilistic risk assessment, NY, IEEE Press, 1992.

[4] McCormick, N.J., Reliability and risk analysis, New York, Academic Press, 1981.

[5] PRA Procedures Guide, Vols. 1&2, NUREG/CR-2300, January 1983.

[6] Probabilistic Risk Assessment Procedures Guide for NASA Managers and Practitioners,NASA, 2002.

[7] Rausand, M. and Hoyland, A., System Reliability Theory, Wiley, 2004.

[8] WASH-1400, Reactor Safety Study, 1975.

9

Estimation of Reliability Parameters from Experimental Data

9.1 Estimation of equipment reliability from tests

To obtain information about the life distribution $F_T(t)$ of a component, it is necessary to carry out a 'life test' where n identical units of the component are activated and their lifetimes recorded.

The fundamental assumptions that are made are that the lifetimes of the n components are statistically independent and identically distributed according to the continuous distribution function $F_T(t)$.

The assumption of identically distributed lifetimes corresponds to the assumption that the components are nominally identical, that is of same type and exposed to approximately the same environmental and operational stresses.

The assumption of independence means that the components are not affected by the operation or failure of any other component in the set.

Any censoring mechanism (see below) must also be 'independent', i.e. censorings occur independent of any information gained from previously failed components in the set.

9.1.1 Complete data set

If the test is allowed to run until all the n components have failed and the lifetimes are recorded, the data set thereby obtained is said to be complete (Figure 9.1).

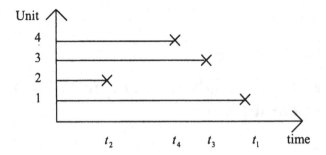

Fig. 9.1: Life timelines for a complete failure data set of $n=4$ test components. The symbol × indicates failure

9.1.2   Censored data sets

Often, it is impractical or too expensive to wait until all the components have failed. Hence, censoring is applied to cease the test before all components have failed.

A <u>right-censored</u> data set is composed also of units that did not fail during the test (Fig. 9.2).

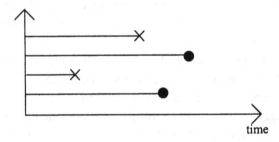

Fig. 9.2: Life timelines for a right-censored data set of $n=4$ test components. The symbol × indicates failure; the symbol • indicates success

An <u>interval-censored</u> data set reflects uncertainty as to when the units actually failed, due to the fact that units are inspected at fixed times so that their statuses are known only at the time of inspection. Thus, the failure of a unit is revealed upon inspection and it is known only that it occurred between inspections, but not the exact time of failure.

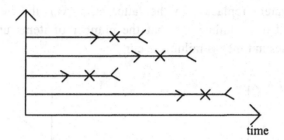

Fig. 9.3: Life timelines for a complete failure data set of $n = 4$ test components. The symbol × indicates failure; the symbols > < indicate the uncertainty interval due to the inspection scheduling

A <u>left-censored</u> data set is a special case of interval censored data in which the time-to- failure for a particular unit is known to occur between time zero and some inspection time.

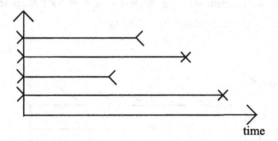

Fig. 9.4: Life timelines for a complete failure data set of $n = 4$ test components. The symbol × indicates failure; the symbols > < indicate the uncertainty interval due to the inspection scheduling

9.1.3 Test plans

Test plans are characterized by (Figs. 9.5 and 9.6):

-   the moment of termination of testing, at fixed time t_0 (Type I) or at the r -th failure (Type II);
-   whether items are replaced upon failure (R) or not (W).

    In the former case (R), n items are under testing at all times: the units are continuously monitored and upon failure they are immediately replaced. In the latter case (W), the items are not replaced upon failure, so that the number of items under testing decreases in time, as failures occur.

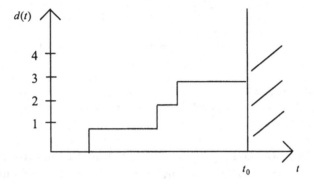

Fig. 9.5: Test I (R or W). The variable $d(t)$ denotes the number of failures that occur before t. The information resulting from the test is: $s \leq n$ observed lifetimes $t_1, t_2, ..., t_s$; $n - s$ components survived up to t_0

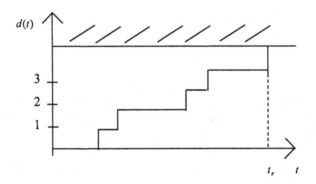

Fig. 9.6: Test II (R or W). $d(t)$ denotes the number of failures that occur before t. The information resulting from the test is: $r \leq n$ observed lifetimes $t_1, t_2, ..., t_r$; $n - r$ components survived up to t_r

There exists also a type III censoring which is a combination of I and II: the test terminates at the time that occurs first, t_0 or t_r.

Finally, if n identical units are activated at different points in time and censored stochastically, the censoring is said to be of type IV.

9.1.4 The method of maximum likelihood estimation applied to test components lifetimes

Consider a data set of uncensored component lifetimes, t_1, t_2, ...,t_n, realizations of the underlying failure time probability density function $f_T(t|\theta)$ where θ is the parameter of the distribution which we wish to estimate. The likelihood of the lifetime realisations observed is:

$$L(\theta) = \prod_{i=1}^{n} f_T(t_i \mid \theta) \tag{9.1}$$

For right-censored data with some components surviving the test, the likelihood function becomes:

$$L(\theta) = \underbrace{\prod_i f_T(t_i \mid \theta)}_{\text{failures}} \underbrace{\prod_j R(t_j \mid \theta)}_{\text{right-censored}} \tag{9.2}$$

where $R(t_j \mid \theta)$ is the reliability of the component at time t_j at which the test of the j-th unit, still functioning, has been interrupted.

To compute an estimate $\hat{\theta}$ of the unknown distribution parameter θ, based on the available data set, generally one takes the log-likelihood

$$l(\theta) = \ln[L(\theta)] \tag{9.3}$$

and maximises with respect to θ:

$$\frac{\partial l}{\partial \theta} = 0 \tag{9.4}$$

As an example, let us consider the right-censored testing of n exponential units with $f_T(t) = \lambda e^{-\lambda t}$. Let r be the number of failure observations. Then, the likelihood is given by

$$L(\lambda) = \lambda^r e^{-\lambda \sum_{i=1}^{r} t_i} e^{-\lambda \sum_{j=r+1}^{n} t_j} = \lambda^r e^{-\lambda \sum_{k=1}^{n} t_k} \tag{9.5}$$

and the log-likelihood is

$$l(\lambda) = r \ln \lambda - \lambda \sum_{k=1}^{n} t_k \tag{9.6}$$

Taking the derivative,

$$\frac{\partial l}{\partial \lambda} = \frac{r}{\lambda} - \sum_{k=1}^{n} t_k \tag{9.7}$$

By setting (9.7) equal to zero, one obtains the estimate $\hat{\lambda}$ of the component failure rate,

$$\hat{\lambda} = \frac{r}{\sum_{k=1}^{n} t_k} = \frac{r}{T} = \frac{\# \, of \, failures \, observed}{total \, test \, time} \tag{9.8}$$

where T is here used to denote the total test time of the component.

9.1.5 Statistics of exponential components with or without replacement

Let us consider in further details the life test of a component characterized by exponentially distributed failure times. This is the most common assumption for reliability and risk calculations. We can use the failure data obtained in the test within a Maximum Likelihood

Estimation (MLE) procedure to estimate the parameter λ of the exponential distribution, i.e. the failure rate of the component. We consider two cases:

W. Without replacement

Given n components under test, one could wait for all of them to fail but that would take a long time for very reliable components. Therefore, one censors the test fixing an end time t_0 (I) or stopping the test at the r-th failure (II).

IW. TYPE I (test ends at t_0); without replacement.

The sample of lifetime data is $(t_1, t_2, ..., t_r, t_0)$, where the first r times are the failure times of the r components which fail within the censoring time t_0.

Then, at t_0 there are still $n - r$ components functioning. The likelihood function $L(\lambda)$ for this case reads:

$$L(\lambda) \propto \left(\lambda e^{-\lambda t_1}\, dt\right)\left(\lambda e^{-\lambda t_2}\, dt\right) \cdots \left(\lambda e^{-\lambda t_r}\, dt\right) e^{-\lambda(n-r)t_0} \tag{9.9}$$

$$L(\lambda) = \lambda^r e^{-\lambda\left[\sum_{j=1}^{r} t_j + (n-r)t_0\right]} = \lambda^r e^{-\lambda T} \tag{9.10}$$

where the unit total time on test is $T = \sum_{j=1}^{r} t_j + (n-r)t_0$.

Setting the derivative with respect to λ equal to zero,

$$\frac{\partial L}{\partial \lambda} = 0, \quad r\lambda^{r-1} e^{-\lambda T} - \lambda^r T e^{-\lambda T} = 0 \tag{9.11}$$

the maximum likelihood estimate of the component failure rate is obtained:

$$\hat{\lambda} = \frac{r}{T} = \frac{\# \, of \, failures \, observed}{total \, test \, time} \tag{9.12}$$

IIW. TYPE II (test ends at r-th failure); without replacement.

The sample of lifetime data is $(t_1, t_2, ..., t_r)$, where the first r times are the failure times of the r components which fail within the censoring time t_r.

Then, at t_r there are still $n - r$ components functioning. The likelihood function $L(\lambda)$ for this case reads:

$$L(\lambda) \propto \left(\lambda e^{-\lambda t_1} \, dt\right)\left(\lambda e^{-\lambda t_2} \, dt\right) \cdots \left(\lambda e^{-\lambda t_r} \, dt\right) e^{-\lambda(n-r)t_r} \tag{9.13}$$

$$L(\lambda) = \lambda^r e^{-\lambda\left[\sum_{j=1}^{r} t_j + (n-r)t_r\right]} = \lambda^r e^{-\lambda T} \tag{9.14}$$

where the unit total time on test is $T = \sum_{j=1}^{r} t_j + (n-r)t_r$.

Setting the derivative with respect to λ equal to zero,

$$\frac{\partial L}{\partial \lambda} = 0, \quad r\lambda^{r-1}e^{-\lambda T} - \lambda^r T e^{-\lambda T} = 0 \tag{9.15}$$

the maximum likelihood estimate of the component failure rate is obtained:

$$\hat{\lambda} = \frac{r}{T} = \frac{\# \, of \, failures \, observed}{total \, test \, time}$$

$$\tag{9.16}$$

R. With replacement

When a test with replacement is considered, every time a component fails it is replaced by a new identical one.

IR. TYPE I (test ends at t_0); with replacement.

Let us consider again the testing of n identical units. At some time t_1 one fails and $n-1$ are left under test. Since the components are exponential (no aging), the failed component can be replaced with a new identical one, so that at any time there are n units under test, whose failure times follow the same failure distribution (Figure 9.7). This is possible only for exponential components, which do not suffer aging. At the generic time t, the total test time is $n \cdot t$.

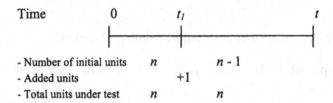

Figure 9.7: Testing with replacement

The sample of lifetime data is $(t_1, t_2, ..., t_r, t_0)$. The likelihood function $L(\lambda)$ reads:

$$L(\lambda) \propto \left(\lambda e^{-\lambda t_1} dt\right)\left(\lambda e^{-\lambda t_2} dt\right)\cdots \lambda e^{-\lambda t_r} \cdot C(\lambda) \qquad (9.17)$$

where $C(\lambda)$ is a complicated expression related to the failure and survival of the replaced units.

For example, for the case of $n = 1$ component that is replaced at each failure, the likelihood is:

$$L(\lambda) = \lambda e^{-\lambda t_1} \lambda e^{-\lambda(t_2-t_1)} \lambda e^{-\lambda(t_3-t_2)} \cdots \lambda e^{-\lambda(t_r-t_{r-1})} e^{-\lambda(t_0-t_r)}$$
$$= \lambda^r e^{-\lambda t_0} \tag{9.18}$$

where $T = t_0$ is the total time on test of a unit of that kind.

For n components,

$$L(\lambda) = \lambda^r e^{-\lambda n t_0} = \lambda^r e^{-\lambda T} \tag{9.19}$$

where $T = n \cdot t_0$ is the total time on test of a unit of that kind. Then, the maximum likelihood estimate is obtained by setting the derivative of $L(\lambda)$ equal to zero:

$$\hat{\lambda} = \frac{r}{T} \tag{9.20}$$

IIR. TYPE II (test ends at r-th failure); with replacement.

The sample of lifetime data is $(t_1, t_2, ..., t_r)$. The likelihood function $L(\lambda)$ then reads:

$$L(\lambda) = \lambda^r e^{-\lambda n t_r} = \lambda^r e^{-\lambda T} \tag{9.21}$$

where the total time on test is $T = n t_r$. Then, the failure rate is estimated by:

$$\hat{\lambda} = \frac{r}{T} \tag{9.22}$$

In general, then, the estimate of the failure rate is always given by the ratio between the number of failures r and the total test time T, *i.e.* $\hat{\lambda} = \frac{r}{T}$, independently of the test strategy, and what changes is the total time on test T depending on the type of test.

9.1.6   Confidence limits for reliability parameters

It is difficult to generalise about a given statistical population when only a point value characteristic (e.g. the mean) of a finite size sample is measured, since such sample may not be representative of the population. As the sample size increases, the characteristic values of the sample and those of the population will, of course, agree more closely.

Example 9.1: Coin toss (binomial)

Consider a simple experiment of coin tossing, giving the following results in two samples of size 10 and 1000 tosses, respectively:

$$4\,heads, 10\,tosses$$

$$400\,heads, 1000\,tosses$$

In both cases the maximum likelihood estimate of the probability of heads can be shown to be:

$$\hat{p}_{MLE} = \frac{\#\ of\ heads}{\#\ of\ tosses} = 0.4$$

The point estimates are the same in the two cases but it is intuitive that there is more confidence in the experimental evidence given by the second sample. Evidently, the point estimates do not give a measure of the confidence in the result.

   Since one cannot be certain that a sample is representative of a population, it is important to associate a 'degree of confidence' to an estimated sample characteristic. In particular, we are interested in associating confidence limits to probabilistic parameters estimated from reliability tests, e.g. the failure rate and the reliability of a component. For example, one would like to be $(1-2\alpha)$ confident that the unknown true reliability of a component is at least (or at most) a certain value, where α, the level of confidence, must lie in [0,0.5]. Figure 9.8 shows a set of one-sided upper limits of the unknown reliability at a given confidence level, estimated form 5 different reliability test samples. Figure 9.9 shows two-sided confidence limits at a given confidence level, for 7 different reliability test samples. Note that the confidence interval varies according to the results of the different tests.

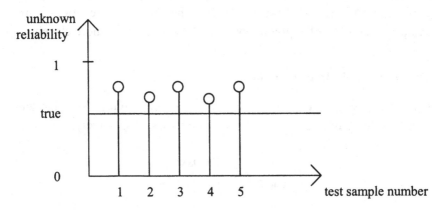

Fig. 9.8: The circles represent the values of the one-sided upper limit at a given confidence level, for different reliability test samples [1]

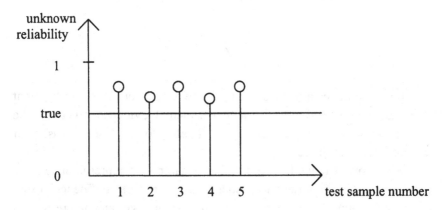

Fig. 9.9: The circles define the upper and the lower limits of the two-sided confidence interval at a given confidence level for different reliability test samples [1]

Consider a sample of realizations $(t_1, t_2, ..., t_n)$ drawn from the population distribution. Let ϑ be the unknown characteristic of the population, e.g. the mean μ or the standard deviation σ and $S = g(t_1, t_2, ..., t_n)$ the corresponding estimator. S is a function of the random sample $(t_1, t_2, ..., t_n)$, with distribution $F_S(s|\vartheta)$ dependent on ϑ. The two-sided confidence interval of S at a level of confidence $1 - 2\alpha$ is obtained by determining the values $s_1(\vartheta)$ and $s_2(\vartheta)$ such that (Figure 9.10):

$$\int_{-\infty}^{s_1(\vartheta)} f_S(s|\vartheta)ds = \alpha \quad ; \quad \int_{s_2(\vartheta)}^{\infty} f_S(s|\vartheta)ds = \alpha \tag{9.23}$$

These are equivalent to

$$Pr[s_1(\vartheta) \le S] = 1 - \alpha \quad ; \quad Pr[S \le s_2(\vartheta)] = 1 - \alpha \tag{9.24}$$

which lead to (Figure 9.10)

$$Pr[s_1(\vartheta) \le S \le s_2(\vartheta)] = 1 - 2\alpha \tag{9.25}$$

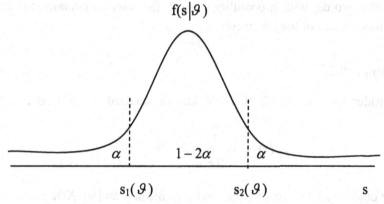

Fig. 9.10: Confidence interval at level $1 - 2\alpha$ [1]

The above expressions can be rewritten to express explicit inequalities in terms of the unknown characteristic ϑ [1].

Two random variables θ_1 and θ_2 are introduced as the following functions of the estimator:

$$\theta_1 = s_1^{-1}(s)$$
$$\theta_2 = s_2^{-1}(s)$$
(9.26)

From (9.24), it follows that

$$P[\vartheta \le \theta_1] = 1 - \alpha$$
$$P[\theta_1 \le \vartheta] = 1 - \alpha$$
(9.27)

Thus, the introduced random variables θ_1 and θ_2 constitute upper and lower $(1-\alpha)$-confidence limits of the unknown parameter ϑ, so that the random interval $[\theta_2, \theta_1]$ becomes the $100(1-2\alpha)\%$ confidence interval, i.e.

$$P[\theta_2 \le \vartheta \le \theta_1] = 1 - 2\alpha$$
(9.28)

In other words, with probability $1 - 2\alpha$ the interval contains the true, unknown value of the parameter ϑ.

Example 9.2

Consider a normal distribution with known standard deviation σ:

$$f_T(t) = \frac{1}{\sqrt{2\pi}\sigma} e^{-\frac{(t-\mu)^2}{2\sigma^2}}$$

The objective is to estimate the unknown mean μ and its 90% confidence interval from a sample of n realizations $(t_1, t_2, ..., t_n)$. This means that $1 - 2\alpha = 0.9$ or $\alpha = 0.05$ and that we must find the 5-th and the 95-th percentiles of the distribution of the estimator.

Solution:

The 90% confidence interval means that $\alpha = 0.9$, from which $1 - \alpha = 0.05$.

The estimate of μ is the sample mean, i.e.:

$$\hat{\mu} = \bar{t} = \frac{\sum_{i=1}^{n} t_i}{n}$$

The random variable $\bar{t} \approx N\left(\mu, \frac{\sigma^2}{n}\right)$. Then, passing to the standard normal variable $\xi \approx N(0,1)$ (Chapter 4),

$$P\left[\xi_{0.5} < \frac{\bar{t} - \mu}{\sigma/\sqrt{n}} < \xi_{0.95}\right] = P\left[-1.645 < \frac{\bar{t} - \mu}{\sigma/\sqrt{n}} < 1.645\right] = 0.90$$

where $\xi_{0.5}$ and $\xi_{0.95}$ represent the 5-th and the 95-th percentiles of the standard normal distribution $N(0,1)$ tabuled in Appendix A.

Notice that this probabilistic statement refers to the estimator $\bar{t}$, which is the random variable in question, not to μ and σ which are the distribution parameters.

Given that σ is known and solving for μ:

$$P\left[\bar{t} - 1.645\frac{\sigma}{\sqrt{n}} < \mu < \bar{t} + 1.645\frac{\sigma}{\sqrt{n}}\right] = 0.9$$

where the interval $\left[\bar{t} - 1.645\frac{\sigma}{\sqrt{n}} ; \bar{t} + 1.645\frac{\sigma}{\sqrt{n}}\right]$ depends on $\bar{t}$ and is, thus, random itself.

For a different random $(t_1, t_2, \ldots, t_n)$, one obtains a different estimate $\bar{t}$ of μ and thus a different confidence interval (Figure 9.11).

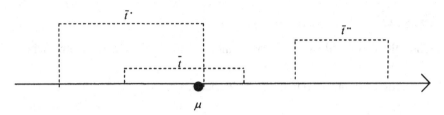

Fig. 9.11: Random confidence intervals for three different samples and corresponding estimates $\bar{t}$, $\bar{t}\,^{\cdot}$ and $\bar{t}\,^{\cdot\cdot}$ of the mean value μ.

Hence, the above 90%-probability statement concerning the confidence interval means that taking 1000 samples and evaluating each time the estimate $\bar{t}$ and the corresponding confidence interval, 900 times out of 1000 the interval will include μ. In other words, the sample average $\bar{t}$ gives an estimate of μ and the corresponding confidence interval contains the unknown μ with a probability of 0.9. Notice that increasing the sample size n leads to tighter intervals, as the dependence goes with $\frac{1}{\sqrt{n}}$. In other words, confidence increases with the sample size.

Example 9.3

Consider a sample of type II-censored exponentially distributed data (fixed number of failures r, Section 9.1.5). The estimate of the failure rate is:

$$\hat{\lambda} = \frac{r}{T} \qquad T = \sum_{j=1}^{r} t_j + (n-r)t_r$$

Find the α-confidence limits of the Mean Time to Failure MTTF $\hat{\vartheta} = \frac{1}{\hat{\lambda}}$.

Solution:

What is the probability distribution of the estimator $\hat{\lambda}$?

Let D_j denote the time interval from the $(j - 1)$-*th* to the j-*th* failure; then,

$$T_1 = D_1$$
$$T_2 = D_1 + D_2$$
$$\ldots..$$
$$T_r = D_1 + D_2 + \ldots\ldots.. + D_r$$

and

$$\sum_{j=1}^{r} T_j = rD_1 + (r-1)D_2 + \ldots\ldots + D_r$$

$$(n - r)T_r = (n - r)(D_1 + D_2 + \ldots + D_r)$$

Therefore, the total test time of the unit at time T_r is:

$$T = nD_1 + (n-1)D_2 + \ldots + [n - (r-1)]D_r = \sum_{j=1}^{r}[n - (j-1)]D_j = \sum_{j=1}^{r} D_j^*$$

It can be proved that $2\lambda D_1^*$, $2\lambda D_2^*$, ..., $2\lambda D_r^*$ are independent and χ^2 - distributed random variables, each one with two degrees of freedom [1], so that $2\lambda T$ is also χ^2 -distributed with $2r$ degrees of freedom. This distribution is tabulated in Appendix B.

Let $\chi_\alpha^2(2r)$ be the 100α percentile of the chi-square distribution, with $0 \le \alpha \le 1$. Then, from the definition of the percentiles:

$$P[2\lambda T \le \chi_\alpha^2(2r)] = \alpha$$

or

$$P[\lambda \le \frac{\chi_\alpha^2(2r)}{2T}] = \alpha$$

From percentiles, the confidence limits for the estimate of MTTF $\hat{\vartheta} = \frac{1}{\hat{\lambda}}$ can be obtained, for both censoring of Type I and II :

	I, fixed t_0		II, fixed r
one-sided (lower)	$\dfrac{2T}{\chi_\alpha^2 \underbrace{(2r+2)}_{\text{\# of degrees of freedom}}}$ $\downarrow$ percentile		$\dfrac{2T}{\chi_\alpha^2(2r)}$
two-sided (lower and upper)	$\dfrac{2T}{\chi_{\frac{1+\alpha}{2}}^2(2r+2)}$,	$\dfrac{2T}{\chi_{\frac{1-\alpha}{2}}^2(2r)}$	$\dfrac{2T}{\chi_{\frac{1+\alpha}{2}}^2(2r)}$, $\dfrac{2T}{\chi_{\frac{1-\alpha}{2}}^2(2r)}$

For example, the one-sided, lower α-confidence limit for the test of type I (fixed t_0), is such that $P\left[\hat{\vartheta} \geq \dfrac{2T}{\chi_\alpha^2(2r+2)} \right] = \alpha$

Example 9.4

Estimate the 95*th* percentile for the MTTF of a nuclear reactor given a sample of 1 failure (Three Miles Island) in 2000 reactor-years. The value of $\chi_{0.95}^2(4)$ is 9.49. Then,

$$\hat{\vartheta}_{0.95} = \frac{2 \cdot 2000}{9.49} = 421 \; reactor\text{-}years$$

Example 9.5

Assume 30 identical components placed on Type II censoring with $r = 20$. The 20th failure has a time to failure (TTF) of 39.89 min, i.e., $t_r = 39.89$, and the other 19 times to failure are listed in Table 9.1 along with times to failure which would occur if the test were to continue after the 20th failure. Find the 95% two-sided confidence limits for the MTTF, $\hat{\vartheta}$.

Solution:

Let $N = 30$, $r = 20$, $t_r = 39.89$, $\alpha = 0.95$.

$$\hat{\vartheta} = S = \frac{(30-20) \times 39.89 + 291.09}{20} = 34.50$$

From the chi-square distribution,

$$\chi^2_{\frac{1-\alpha}{2}}(2r) = \chi^2_{0.025}(40) = 59.3417$$

$$\chi^2_{\frac{1+\alpha}{2}}(2r) = \chi^2_{0.975}(40) = 24.4331$$

Then,

$$\Theta_1 = \frac{2T}{\chi^2_{\frac{1+\alpha}{2}}(2r)} = 2r \cdot \frac{\hat{\vartheta}}{\chi^2_{0.975}(2r)} = 2 \times 20 \times \frac{34.50}{59.3417} = 23.26$$

$$\Theta_2 = \frac{2T}{\chi^2_{\frac{1-\alpha}{2}}(2r)} = 2r \cdot \frac{\hat{\vartheta}}{\chi^2_{0.025}(2r)} = 2 \times 20 \times \frac{34.50}{24.4331} = 56.48$$

Then,

$$23.36 \leq \vartheta \leq 56.48$$

Thus, we are 95% confident that the true, unknown mean time to failure (ϑ) is in the interval [23.36, 56.48]. As a matter of fact, TTFs in Table 9.1 were generated from an exponential distribution with the MTTF, $\vartheta = 26.64$. The obtained confidence interval includes this true MTTF.

Table 9.1: TTFs Data for Example 9.5

TTFs Up to 20th Failure				TTFs After 20th Failure	
t_1	0.26	t_{11}	11.04	t_{21}	(40.84)
t_2	1.49	t_{12}	12.07	t_{22}	(47.02)
t_3	3.65	t_{13}	13.61	t_{23}	(54.75)
t_4	4.25	t_{14}	15.07	t_{24}	(61.08)
t_5	5.43	t_{15}	19.28	t_{25}	(64.36)
t_6	6.97	t_{16}	24.04	t_{26}	(64.45)
t_7	8.09	t_{17}	26.16	t_{27}	(65.92)
t_8	9.47	t_{18}	31.15	t_{28}	(70.82)
t_9	10.18	t_{19}	38.70	t_{29}	(97.32)
t_{10}	10.29	t_{20}	39.89	t_{30}	(164.26)

For the reliability of this kind of component, with exponential distribution (Chapter 4):

$$R(t) = e^{-\lambda t} = e^{-t/\vartheta}$$

and the confidence intervals can be obtained by substituting Θ_1 and Θ_2:

$$e^{-t/\Theta_2} \le R(t) \le e^{-t/\Theta_1}$$

Thus, for the data in Example 2,

$$e^{-t/23.36} \le R(t) \le e^{-t/56.48}$$

Similarly, the confidence interval for the true, unknown failure rate λ is given by

$$\frac{1}{\Theta_1} \le \lambda \le \frac{1}{\Theta_2}$$

$$\frac{1}{56.48} \le \lambda \le \frac{1}{23.36}$$

9.2   Accelerated Life Testing

9.2.1   Introduction

Many of the devices produced today for complex technical systems have very high reliability under normal use conditions. The time involved in a life test such as those described in Section 9.1 would therefore be prohibitive. Furthermore, the device is likely to be obsolete by the time the test is completed. The questions then arise of how to make the optimal choice between several types or designs of a device and how to collect information about the corresponding life distributions under normal use conditions.

A common way of tackling these problems is to expose the device to sufficient overstress to bring the mean time to failure down to an acceptable level. Thereafter, one tries to "extrapolate" from the information obtained under over stress to normal use conditions. This approach is called *Accelerated Life Testing (ALT)* or overstress testing [2-8, 11].

Depending on the kind of device in question, the accelerated testing conditions may involve a higher level of temperature, pressure, voltage, load, vibration, and so on, than the corresponding levels occurring in normal use conditions. These variables are called *stressors* (stress variables or covariates). In a specific situation, there may be one or several (*m*) stressors $s_1, s_2, ..., s_m$ acting simultaneously. The vector $s = (s_1, s_2, ..., s_m)$ is called the *stress vector*.

In simple situations, there is only one stressor s occurring on two levels $s^{(1)}$ and $s^{(2)}$, where $s^{(1)} < s^{(2)}$. Let $s^{(0)} \leq s^{(1)}$ denote normal stress. The situation becomes somewhat complicated when m stressors $s_1, s_2, ..., s_m$ are involved and stressor s_j occurs on n_j levels,

$$s_j^{(1)} < s_j^{(2)} < s_j^{(n_j)} \text{ for } j = 1, 2, ..., m \qquad (9.29)$$

Let $s_j^{(0)}$ ($\leq s_j^{(1)}$) denote normal stress for stressor j, for $j = 1, 2, ..., m$. The situation becomes more complicated when the stressors are continuously increasing with time. The first two cases lead to *Step-stress Accelerated*

Life Tests (SALTs); the last one leads to *Progressive-stress Accelerated Life Tests* (PALT).

9.2.2   Experimental designs for ALT

Let us for the sake of simplicity suppose that there is only *one* stressor s. The testing experiment can be conducted according to different designs. We will discuss three such designs.

Design I

The experiment involves use of k stress levels $s^{(1)} < s^{(2)} < ... < s^{(k)}$ (see Figure 9.12). Let $s^{(0)} < s^{(1)}$ denote normal stress. A (large) number of test units are assumed to be available for the experiment and n_j of these are to be exposed to the stress $s^{(j)}$. Censoring of type II (test terminates at r-th failure, Section 9.1.3) is applied. The experiment is then carried out as follows:

1.   One stress level $s^{(i)}$ is *chosen at random* among $s^{(1)}$, $s^{(2)}$,..., $s^{(k)}$ and n_i test units are chosen at random among the test units at hand. These n_i units are then exposed to stress level $s^{(i)}$. The test is terminated when r_i ($\leq n_i$) failures have occurred. Let $T_{i1}, T_{i2},...,T_{in_i}$, denote the times to failure or censoring.

2.   Another stress level $s^{(j)}$ is chosen at random among the remaining levels, n_j test units are chosen at random among the remaining units and exposed to stress level $s^{(j)}$. The test is terminated when $r_j \leq n_j$ failures have occurred. Let $T_{j1}, T_{j2},...,T_{jn_j}$ denote the times to failure or censoring. This procedure is continued until k stress levels have been selected.

Figure 9.12: Design I for accelerated tests

If the number of test units at hand is large compared to $n = \sum_{j=1}^{k} n_j$, it seems reasonable to assume that $T_{01}, T_{02}, ..., T_{kr_k}$ are independent, which simplifies the analysis.

Design II

Fix k points of time $0 < t_1 < t_2 < ... < t_k < t$. Put n randomly chosen test units on test at time 0. In the time interval $(0, t_1]$ the units are subject to stress $s^{(1)}$; in the interval $(t_1, t_2]$ the units that have not failed by time t_1 are kept in operation under stress $s^{(2)}$. In the next interval $(t_2, t_3]$ the units that still have not failed by time t_2 are kept in operation under stress $s^{(3)}$, and so on (see Figure 9.13). In the time interval $(t_k, \infty]$ the units that have not failed by time t_k are kept in operation under stress $s^{(k+1)}$ until they have all failed (hence, no censoring). The lifetimes of the n test units are denoted $T_1, ..., T_n$.

Stress level

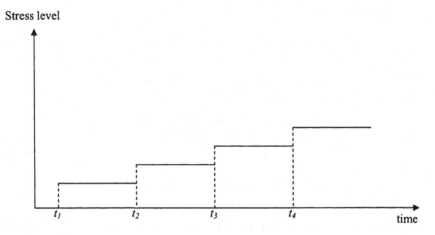

Figure 9.13: Design II for accelerated tests

Design III

A number n of test units are chosen at random among the test units at hand and exposed to a stress $s(t)$, which is increasing with time until the units have all failed. The stress function $s(t)$ is assumed known (Figure 9.14). The lifetimes of the n test units observed are denoted $T_1, T_2, ..., T_n$.

If n is small compared to the number of units at hand and if the n units are operating independently, it seems reasonable to assume that $T_1, ..., T_n$ are independent, in both design II and design III.

Stress level

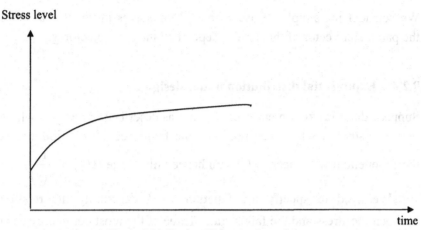

Figure 9.14: Design III for accelerated tests

9.2.3   Parametric models used in step-stress accelerated tests

The data obtained in the test is supposed to give information about:

–   the lifetime distribution function $F_T(t;\underline{s}) = P(T \leq t;\underline{s})$

–   the survival function $R_T(t;\underline{s}) = 1 - F_T(t;\underline{s})$

–   the failure rate $\lambda(t;\underline{s}) = \dfrac{f_T(t;\underline{s})}{1 - F_T(t;\underline{s})}$

For the sake of simplicity, let us suppose that we succeed in establishing an a priori, parametric life distribution under normal use conditions, e.g. exponential or Weibull etc. What will be the effect of overstress on this baseline distribution?

     There are two alternatives:

1.   Different stress levels only lead to different parameters values but leave unchanged the form of the distribution
2.   Different stress levels modify also the type of distribution.

We consider the simplest Case 1 and ask ourselves in which way does the parameter vector of this family depend on the stress vector $\underline{s}$.

9.2.4 Exponential distribution under design I

Suppose that the experiment is carried out as described in design I where only one stressor s has been used and the family of life distributions is the exponential with mean $\vartheta\,(s)$, and hence failure rate $\lambda(s) = \dfrac{1}{\vartheta(s)}$.

We need to specify the function $\lambda(s)$ describing the relation between the stress and the failure rate. Three of the most commonly used relations are:

$$\vartheta(s) = c \cdot s^{-a} \qquad \begin{array}{c}\text{power rule model (dielectric breakdown of}\\ \text{capacitors and fatigue testing of materials)}\end{array} \qquad (9.30)$$

$$\lambda(s) = c \cdot e^{-b/s} \qquad \begin{array}{c}\text{Arrhenius model (thermal aging and}\\ \text{semiconductor materials)}\end{array} \qquad (9.31)$$

$$\lambda(s) = c \cdot e \cdot s^{-b/s} \qquad \text{simple Eyring model (constant thermal stress)} \quad (9.32)$$

The constants a, b, c have to be estimated on the basis of the recorded life lengths under overstress. Inserting the expression for $\lambda(s)$ in the underlying (exponential) lifetime distribution, $F_T\left(t; s^j\right)$ is now known for any stress level $s^{(j)}$, $j = 1, 2,..., k$, except for the values of the constants a, b, c which could be estimated from data by applying, for example, the maximum likelihood estimation (MLE) or least squares (LS) methods. The estimate $\hat{a}, \hat{b}, \hat{c}$ can then be inserted, together with the normal stress level $s^{(0)}$ in the lifetime distribution $\hat{F}_T\left(t; s^{(0)}\right)$.

The total information that is obtained through the accelerated life test under design I is:

$s^{(j)}$ = stress level

n_j = number of units tested at stress level $s^{(j)}$ $\qquad\qquad j = 1, 2, ..., k$

r_j = number of units failed under $s^{(j)}$

$T_{j1}, T_{j2}, ..., T_{jn_j}$ = lifetimes of the n_j tested units which can be combined in

$$T_j = \sum_{i=1}^{r_j} T_{ji} + (n_j - r_j) T_{jr_j} = \text{total time on test.}$$

Then, we know that for type II-censored, exponentially-distributed data, the variable $Z_j = 2\lambda(s^{(j)})T_j$ is χ^2-distributed with $2r_j$ degrees of freedom, $j = 1, 2, ..., k$:

$$f_{Z_j}(z_j) = \frac{1}{2^{r_j}\,\Gamma(r_j)} z_j^{r_j} e^{-\frac{z_j}{2}} \qquad z_j > 0, j = 1, 2, ..., k \qquad (9.33)$$

Accordingly, $f_{Z_j}(z_j)dz_j = 2\lambda(s^{(j)})f_{T_j}(t_j)dt_j$ and

$$f_{T_j}(t_j) = \frac{1}{2^{r_j}\,\Gamma(r_j)}\left[2\lambda(s^{(j)})\cdot t_j\right]^{r_j-1}\cdot e^{-\lambda(s^{(j)})t_j}\cdot 2\lambda(s^{(j)}) =$$

$$= \frac{1}{\Gamma(r_j)}\lambda(s^{(j)})^{r_j} t_j^{r_j-1}\cdot e^{-\lambda(s^{(j)})t_j} \qquad t_j > 0, j = 1, 2, ..., k$$

$$(9.34)$$

Hence, the joint distribution of the set $(T_1, T_2, ..., T_k)$ would be:

$$f_{T_1, T_2, ..., T_k}(t_1, t_2, ..., t_k) = \prod_{j=1}^{k}\frac{1}{\Gamma(r_j)}\lambda(s^{(j)})^{r_j} t_j^{r_j-1}\cdot e^{-\lambda(s^{(j)})t_j} \qquad t_j > 0$$

$$(9.35)$$

As an example, let us consider the case where the relation between the stressors and the mean time $\vartheta(s)$ is described by the power rule model for fatigue testing materials (9.30):

$$\vartheta\left(s^{(j)}\right) = \gamma\left(s^{(j)}\right)^{-a} \qquad j = 1, 2, ..., k \tag{9.36}$$

Then,

$$\lambda\left(s^{(j)}\right) = \frac{1}{\gamma}\left(s^{(j)}\right)^{a} \tag{9.37}$$

If we change the power rule slightly, without changing its basic character, to

$$\lambda\left(s^{(j)}\right) = \frac{1}{c}\left(\frac{s^{(j)}}{\tilde{s}}\right)^{a} \tag{9.38}$$

$$\tilde{s} = \prod_{j=1}^{k} s^{(j)} {}^{\displaystyle r_j \Big/ \sum_{i=1}^{k} \eta} \qquad \text{(weighted geometric mean of the } s_j\text{'s)} \tag{9.39}$$

it turns out that the MLE, $\hat{a}$ and $\hat{c}$, of a and c, become asymptotically independent.

Inserting (9.38) into (9.35), we can write the likelihood function, dependent on the unknown parameters a, c for a given sample data $(t_1, t_2, ..., t_k)$:

$$L(a, c; t_1, t_2, ..., t_k) = \prod_{j=1}^{k} \frac{1}{\Gamma(r_j)}\left[\frac{1}{c}\left(\frac{s^{(j)}}{\tilde{s}}\right)^{a}\right]^{r_j} t_j^{r_j - 1} \cdot e^{-\left(\frac{s^{(j)}}{\tilde{s}}\right)^{a}\frac{t_j}{c}} \tag{9.40}$$

and the log-likelihood function:

$$l(a, c; t1, t2, ..., tk) = \sum_{j=1}^{k}\left\{-\ln\Gamma(r_j) - r_j \ln c + ar_j \ln\left(\frac{s^{(j)}}{\tilde{s}}\right) + (r_j - 1)\ln t_j - \frac{1}{c}\left(\frac{s^{(j)}}{\tilde{s}}\right)^{a} t_j\right\} \tag{9.41}$$

The estimates $\hat{a}$ and $\hat{c}$ are obtained by solving the two derivative equations, with respect to a and c:

$$\frac{\partial l}{\partial a} = \sum_{j=1}^{k} -r_j \left(\ln s^{(j)} - \ln \tilde{s} \right) - \sum_{j=1}^{k} \frac{1}{c} \left(\frac{s^{(j)}}{\tilde{s}} \right)^a \ln \left(\frac{s^{(j)}}{\tilde{s}} \right) t_j = 0 \qquad (9.42)$$

$$\frac{\partial l}{\partial a} = \sum_{j=1}^{k} -\frac{r_j}{c} + \sum_{j=1}^{k} \left(\frac{s^{(j)}}{\tilde{s}} \right)^a t_j \frac{1}{c^2} = 0 \qquad (9.43)$$

From (9.39) we have,

$$\ln \tilde{s} = \sum_{j=1}^{k} \frac{r_j}{\sum\limits_{i=1}^{k} r_i} \ln s^{(j)} \qquad (9.44)$$

Thus,

$$\sum_{j=1}^{k} r_j \left(\ln s^{(j)} - \ln \tilde{s} \right) = 0 \qquad (9.45)$$

and (9.42) becomes $\sum\limits_{j=1}^{k} \left(\frac{s^{(j)}}{\tilde{s}} \right)^a \ln \left(\frac{s^{(j)}}{\tilde{s}} \right) \cdot t_j = 0$ from which $\hat{a}$ can be obtained.

Then, from (9.43), we can determine $\hat{c}$:

$$\hat{c} = \frac{1}{\sum\limits_{i=1}^{k} r_i} \sum_{j=1}^{k} \left(\frac{s^{(j)}}{\tilde{s}} \right)^{\hat{a}} \cdot t_j \qquad (9.46)$$

It can then be shown that the asymptotic variances of $\hat{a}$ and $\hat{c}$ are

$$\text{var}_\infty(\hat{a}) = \left[\sum_{j=1}^{k} r_j \left(\ln \frac{s^{(j)}}{\tilde{s}}\right)^2\right]^{-1} \tag{9.47}$$

$$\text{var}_\infty(\hat{c}) = c^2 \left[\sum_{j=1}^{k} r_j\right] \tag{9.48}$$

$$\text{cov}_\infty(\hat{a}, \hat{c}) = 0 \qquad \text{(independence)} \tag{9.49}$$

Finally, the estimate of the failure rate under normal stress $s^{(0)}$ can be computed as,

$$\hat{\lambda}_0 = \frac{1}{\hat{c}} \left(\frac{s^{(0)}}{\tilde{s}}\right)^{\hat{a}} \tag{9.50}$$

9.2.5   Inverse gaussian fatigue failure time distribution under design III

Let us consider n independent units put on test at time 0. In [0, t], the units are subject to normal stress $s^{(0)}$; in [0, ∞] the units that have not failed by time t are kept in operation under stress $s^{(1)} > s^{(0)}$ until they all fail.

We suppose that the accumulated fatigue in the material is modeled as a Wiener process $W_0(y), y \geq 0$, with drift $\eta > 0$ and diffusion parameter $\delta^2 > 0$. The Wiener process $W_0(y)$ is defined to be an independent increment Gaussian process with $W_0(0) = 0$ and mean $E[W_0(y)] = \eta y$, in which each increment $W_0(y_2) - W_0(y_1)$ has variance $\delta^2(y_2 - y_1)$. Failure occurs when the Wiener process $W_0(y)$ crosses a critical boundary ω.

Defining the fatigue time Y as the first time that the fatigue process $W_0(y)$ crosses the critical boundary ω and setting $\mu = \omega/\lambda$, $\lambda = \omega^3/\sigma^2$, then Y has an inverse Gaussian distribution

$$f_Y(y; \mu, \lambda) = \frac{\lambda}{\sqrt{2\pi y^3}} \cdot e^{-\frac{\lambda}{\mu^2}\left[\frac{(y-\mu)^2}{y}\right]} \qquad y > 0, \ \mu > 0, \ \lambda > 0 \qquad (9.51)$$

We now assume that at time t, the stress is changed from $s^{(0)}$ to $s^{(1)}$ and correspondingly the Wiener process changes from $W_0(y)$ to $W_1(y) = W_0[t + \alpha(y - t)]$, (Fig. 9.15).

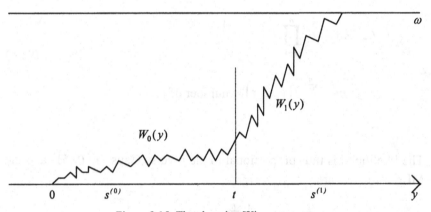

Figure 9.15: The changing Wiener process

It can be shown that in this situation the distribution $F_T(t)$ of the stress failure time T is

$$F_T(y) = \begin{cases} F_0(y) & 0 \le y \le t \\ \\ F_0(t + \alpha(y - t)) & y > t \end{cases} \qquad (9.52)$$

where $F_0(y)$ is the cumulative distribution of the inverse Gaussian density (9.40).

Suppose now that at t, $s^{(0)}$ is changed to $c \cdot s^{(0)}$, with $c > 1$ being a known constant. In this situation it might be reasonable to model the fatigue process $W(y)$ as having drift η in $[0, t)$ and $c \cdot \eta$ in (t, ∞). This means that $\alpha = c$, known.

Let $y_1, y_2, \ldots, y_n$ be the observed failure times:

$$y_j(\alpha) = \begin{cases} y_j & \text{for } y_j \le t \\ \\ t + \alpha(y_j - t)) & \text{for } y_j > t \end{cases} \tag{9.53}$$

In this case of α known, the likelihood function is:

$$L_\alpha(\mu, \lambda) = \alpha^m \prod_{j=1}^{n} f_0 \big[y_j(\alpha)\big] \tag{9.54}$$

$$m = \sum_{j=1}^{n} I\big(y_j > t\big) = \text{number of } y\text{' s} > t$$

The likelihood is thus proportional to the inverse Gaussian (9.51) and the MLE estimates are:

$$\hat{\mu}_\alpha = \frac{1}{n}\sum_{j=1}^{n} y_j(\alpha) \tag{9.55}$$

$$\frac{1}{\hat{\lambda}_\alpha} = \frac{1}{n}\sum_{j=1}^{n}\left[\frac{1}{y_j(\alpha)} - \frac{1}{\hat{\mu}(\alpha)}\right] \tag{9.56}$$

which can be inserted in (9.51) to give an estimate of the life distribution of the failure fatigue time under normal stress $s^{(0)}$.

9.3 Empirical determination of distribution models

Sometimes, the properties of the physical stochastic process under analysis suggest the form of the underlying probability distribution. For example, if a process is composed of the sum of many individual effects, the gaussian distribution may be appropriate on the basis of the central limit theorem. Nevertheless, there are occasions when the required probability distribution has to be determined empirically, that is based solely on the available data.

In practice, the functional form of the probability distribution underpinning a given process is often not easy to derive. Furthermore, an assumed probability distribution (developed theoretically or determined empirically) may be confirmed, or disapproved, in the light of available data using certain statistical tests, known as 'goodness-of-fit' tests.

9.3.1 Probability paper

The simplest and longest used method for parameter estimation is that of probability plotting. This methodology involves plotting the failure times on a specifically-constructed plotting paper to determine the fit of the data to a given distribution and, if applicable, estimates of the distribution parameters.

Graph papers for plotting observed experimental data and their corresponding cumulative frequencies are called 'probability papers'. Probability papers are constructed such that a given probability paper is associated with a specific probability distribution.

Preferably, a probability paper should be constructed using a transformed probability scale in such a manner as to obtain a linear graph between the cumulative probabilities of the underlying distribution and the corresponding values of the variate. Then, the linearity, or lack of linearity, of a set of sample data plotted on a particular probability paper can be used as a basis for determining whether the distribution of the underlying population is the same as that of the probability paper.

Experimental data may be plotted on any probabilistic paper; the 'plotting position' of each data point can be determined as follows:

1. arrange the N observations $x_1, x_2, \ldots, x_N$ in increasing order

2. plot $\left(x_m, \dfrac{m}{N+1} \right)$

9.3.2   The normal probability paper

Let us report on the ordinate axis of a graph the values of the variate X in arithmetic scale and on the abscissa axis two parallel scales: one represents the values of the standard normal variate s whereas the other shows the cumulative probabilities $\Phi_S(s)$, as shown in Fig. 9.16.

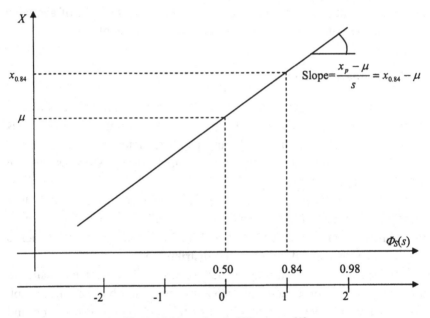

Fig. 9.16: Normal probability paper [9]

A normal value $X \sim (\mu, \sigma)$ would then be represented on this paper by a straight line passing through ($\Phi_S(s) = 0.50$, $X = \mu$) with a slope $\dfrac{x_p - \mu}{s} = \sigma$, where x_p is the value of the variate at probability p. In particular, at $p = 0.84$, $s = 1$ and the slope is then $x_{.84} - \mu$.

If for a given set of data points the resulting graph shows a lack of linearity, this would suggest that the underlying population is not a gaussian, and vice versa.

The mean value and the standard deviation of the underlying population may also be determined graphically from this straight line:

μ_X = the value of X corresponding to $\Phi_S(s) = 0.50$

σ_X = slope of the line $\cong x_{0.84} - \mu$.

9.3.3 The log-normal probability paper

The log-normal probability paper can be obtained from the normal probability paper by simply changing the arithmetic scale for values of the variate X (on the normal probability paper) to a logarithmic scale (Figure 9.17). In this case, the standard normal variate becomes

$$ s = \frac{\ln\left(\dfrac{X}{x_m}\right)}{\xi} \qquad x_m = median\ of\ X \tag{9.57} $$

Accordingly, the median x_m is simply the value of the variate on this line corresponding to $\Phi_s(s) = 0.50$ whereas the parameter ξ is given by the slope of the line, i.e.

$$ \xi = \frac{1}{s}\ln\left(\frac{x_p}{x_m}\right) = \ln\left(\frac{x_{.84}}{x_m}\right) \tag{9.58} $$

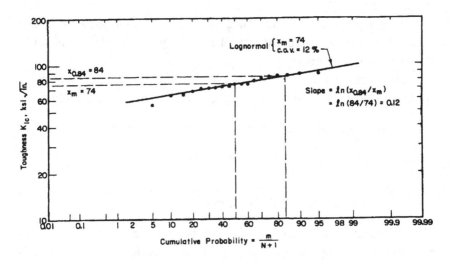

Fig. 9.17: Log-normal probability paper [9]

9.3.4   Construction of a probability plotting paper

As we have seen a probability plotting paper is constructed by linearizing the cumulative density function of the distribution. As an example we will use the well-known Weibull distribution. The cdf of the two-parameter Weibull distribution is.

$$F_T(t) = 1 - e^{-\left(\frac{t}{\tau}\right)^{\beta}} \tag{9.59}$$

We need to linearize this function into the form $y = mx + b$:

$$ln[1 - F_T(t)] = ln\left[e^{-\left(\frac{t}{\tau}\right)^{\beta}}\right] = -\left(\frac{t}{\tau}\right)^{\beta}$$

$$ln\{-ln[1 - F_T(t)]\} = \beta \, ln\left(\frac{t}{\tau}\right)$$

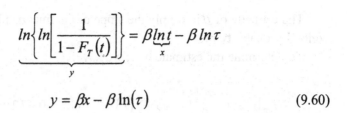

$$y = \beta x - \beta \ln(\tau) \qquad (9.60)$$

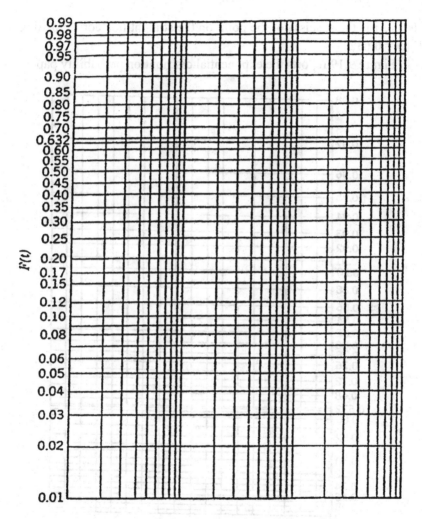

Fig. 9.18: Weibull probability distribution paper [9]

The estimate of β is simply the slope of the linearized line on the Weibull probability plot.

To determine the estimate of τ, we have:

$$F_T(t) = 1 - e^{-\left(\frac{t}{\tau}\right)^\beta} \;\to\; F_T(\tau) = 1 - e^{-1} = 0.632$$

Hence, τ is the abscissa of the point on the straight line corresponding to $F_T(t) = 0.632$.

Figure 9.19 reports the exponential distribution probability paper.

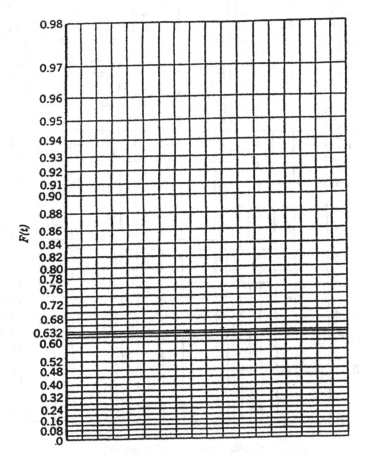

Fig. 9.19: Exponential distribution probability paper [9]

9.3.5   Testing the validity of an assumed distribution

When a theoretical distribution has been assumed, e.g. on the basis of the data plotted on the corresponding probability paper, the validity of the assumed distribution may be verified or disproved statistically by goodness-of-fit tests.

9.3.5.1   Chi-square test

Consider a sample of n observed values of a random variable. The chi-square goodness-of-fit test compares the observed frequencies $n_1, n_2, \ldots, n_K$ of k values of the variate with the corresponding frequencies $e_1, e_2, \ldots, e_k$ from the assumed theoretical distribution. More precisely, we consider the distribution of the quantity:

$$\sum_{i=1}^{k} \frac{(n_i - e_i)^2}{e_i} \qquad (9.61)$$

which approaches the chi-square distribution with $k-1$ degrees of freedom as $n \to \infty$. However, if the parameters of the theoretical model are unknown and must be estimated from the data, the above statement remains valid if the degree of freedom is reduced by one for every unknown parameter that must be estimated.

Let $c_{1-\alpha,\, f}$ be the random variable value corresponding to the cumulative probability value $(1-\alpha)$ of the appropriate χ_f^2 distribution with f degrees of freedom (Figure 9.20). Then, an assumed theoretical distribution is acceptable at the 'significance level α' if

$$\sum_{i=1}^{k} \frac{(n_i - e_i)^2}{e_i} < c_{1-\alpha,f} \qquad (9.62)$$

The basis for this is that if $\displaystyle\sum_{i=1}^{k}\frac{(n_i-e_i)^2}{e_i}$ is a random variable

approximately distributed as χ_f^2, its value has a probability α of being

less than $c_{1-\alpha,f}$.

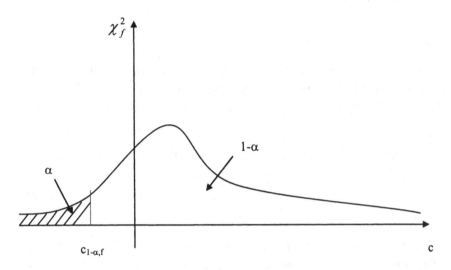

Figure 9.20: Chi-square distribution

In general, the χ^2 - test for goodness of fit gives satisfactory results for
$k \geq 5$, $e_i \geq 5$. Because of the arbitrariness in the choice of the significance
level α, the χ^2-test may not provide absolute information on the
validity of a specific distribution; a distribution may be acceptable at one
significance level α_1 but unacceptable at another one, α_2.

9.3.5.2   Kolmogorov-Smirnov test

The basic procedure for this test involves the comparison between the
experimental cumulative frequency and an assumed theoretical
distribution function.

   If the discrepancy is large with respect to what is normally expected
from a given sample size, the theoretical model is rejected.

For a sample of size n, we rearrange the set of observed data in increasing order and construct the empirical cumulative distribution function (Fig. 9.21):

$$S_n(x) = \begin{cases} 0 & x < x_1 \\ \dfrac{k}{n} & x_k \leq x < x_{k+1} \\ 1 & x \geq x_n \end{cases}$$

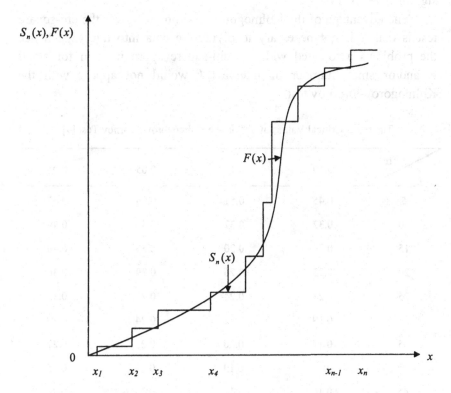

Fig. 9.21: Empirical cumulative frequency vs. theoretical distribution function

In the Kolmogorov-Smirnov test, the maximum difference $D_n = \max\limits_{x} |F(x) - S_n(x)|$ between $S_n(x)$ and $F(x)$ over the entire range of X, is taken as the measure of discrepancy between the theoretical model and the observed data. Theoretically, D_n is a random variable whose distribution depends on n. We can define a critical value D_n^{α} as:

$$P[D_n \leq D_n^{\alpha}] = 1 - \alpha \qquad (9.63)$$

Critical values D_n^{α} at various significance levels α are tabulated in Table 9.2 for various values of n [9].

If $D_n < D_n^{\alpha}$, the proposed distribution is acceptable at the specified significance level α.

The advantage of the Kolmogorov-Smirnov test over the chi-square test is that it is not necessary to divide the data into intervals; hence the problems associated with the chi-square approximation for small e_i and/or small number of intervals k would not appear with the Kolmogorov-Smirnov test.

Table 9.2: Critical Values of D_n^{α} in the Kolmogorov-Smirnov Test [9]

n \ α	0.20	0.10	0.05	0.01
5	0.45	0.51	0.56	0.67
10	0.32	0.37	0.41	0.49
15	0.27	0.30	0.34	0.40
20	0.23	0.26	0.29	0.36
25	0.21	0.24	0.27	0.32
30	0.19	0.22	0.24	0.29
35	0.18	0.20	0.23	0.27
40	0.17	0.19	0.21	0.25
45	0.16	0.18	0.20	0.24
50	0.15	0.17	0.19	0.23
>50	$\dfrac{1.07}{\sqrt{n}}$	$\dfrac{1.22}{\sqrt{n}}$	$\dfrac{1.36}{\sqrt{n}}$	$\dfrac{1.63}{\sqrt{n}}$

9.4   Kaplan-Meier estimator of the survivor function

Let $F_T(t)$ denote the life distribution for a certain type of units. We know the distribution to be continuous, but make no further assumption about $F_T(t)$, i.e. a non parametric model.

Let t_j denote the observed lifetime of unit j. On the basis of the observed lifetimes of n units, $j = 1, 2, ..., n$ we want to estimate the survival function,

$$R(t) = 1 - F_T(t)$$

Then, the empirical cumulative distribution function is (Figure 9.22)

$$\hat{F}_n(t) = \frac{number\ of\ lifetimes \leq t}{n} \qquad (9.64)$$

and the empirical reliability survival function (Figure 9.23)

$$\hat{R}_n(t) = 1 - \hat{F}_n(t) = \frac{number\ of\ lifetimes > t}{n} \qquad (9.65)$$

which is a step function decreasing by $1/n$ at each observed failure time.

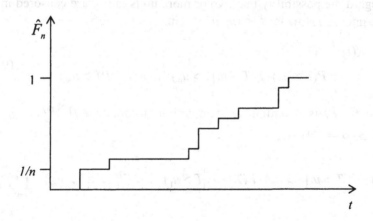

Figure 9.22: Empirical cumulative distribution function

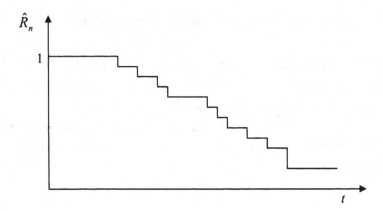

Figure 9.23: Empirical survival function

The Kaplan-Meier estimator is regarded as the most direct non-parametric estimator of the survival function. It is the only coherent estimator of the survival function for censored tests [11].

The basic principle of the estimator is that being in good working condition after t means *i)* being so already before t and *ii)* not failing at t.

Let the time period $[0, \infty]$ be divided into small intervals $(u_j, u_{j+1}]$ for $j = 1, 2, ..., n$, with $u_0 = 0$ and the intervals short enough that we can disregard the possibility that two or more units fail or are censored in the same interval. Now let $t \in (u_j, u_{j+1}]$. Then,

$$R(t) = P(T > t)$$
$$= P(T > u_0) \cdot P(T > u_1 | T > u_0) \cdot .... \cdot P(T > t | T > u_m) \tag{9.66}$$

Since $F_T(t)$ is a continuous life distribution for all $t \geq 0$, $P(T > u_0) = P(T > 0) = 1$. Hence,

$$R(t) = P(T > u_1 | T > u_0) \cdot P(T > u_2 | T > u_1) \cdot ...... \cdot P(T > t | T > u_m) = \prod_{j=0}^{m} p_j \tag{9.67}$$

where $p_j = P(T > u_{j+1} | T > u_j)$ $j = 0, 1, 2, ..., m-1$

$$p_m = P(T > t | T > u_m) \hspace{3cm} (9.68)$$

Kaplan-Meier's idea is then that of estimating each single factor on the right-hand side of (9.51) and thereafter use the product of these estimators as an estimator of $R(t)$.

The estimation procedure follows the steps reported below [11]:

1. If neither failure nor censoring occurs in $(u_j, u_{j+1}]$, then the same number of units will be active at the start and at the end of this interval. Then,

$$\hat{p}_j = 1$$

2. Suppose that censoring of one unit occurs in $(u_j, u_{j+1}]$. Then, due to the assumption of short intervals, we may ignore the possibility that another censoring or failure occurs in the same interval. Accordingly, we record no failures in the intervals and

$$\hat{p}_j = 1$$

3. Suppose that failures occur in $(u_j, u_{j+1}]$. Due to the assumption of short intervals we may ignore the possibility of more than one failure occurring in this interval. Let n_j denote the number of units at risk (i.e. which are functioning and in observation) at the beginning of the interval. The number of units at risk at the end is $n_j - 1$. Then,

$$\hat{p}_j = \frac{n_j - 1}{n_j}$$

Thus, the only intervals where the estimator $\hat{p}_j \neq 1$ are those in which failure occurs. By increasing the number of intervals such that their length approaches zero (except for the last one which goes to ∞), we see that the estimator $\hat{p}_j \neq 1$ only at the failure times, just as in the empirical $\hat{R}$.

From the above, it follows that we may disregard the intervals where no failures occur. We then redefine (to simplify the notation):

n_j = number of units at risk (functioning and in observation) immediately before time t_j, $j = 1, 2, ..., n$.

The probability p_j may now be estimated for infinitesimal intervals around the t_j's:

$$\hat{p}_j = \begin{cases} 1 & \text{if a censoring occured at } t_j \\ \dfrac{n_j - 1}{n_j} & \text{if a failure occured at time } t_j \end{cases} \qquad j = 1, 2, ..., n$$

$$\hat{p}_0 = 1$$

Then, we have the Kaplan-Meier estimator

$$\hat{R}(t) = \prod_{j=0}^{n} \hat{p}_j = \prod_{j \in \Gamma} \frac{n_j - 1}{n_j} \qquad \Gamma = \left\{ t_j \text{ of failure, } t_j < t \right\} \qquad (9.69)$$

The following properties of the Kaplan-Meier estimator are of relevance:

1. It can be derived as a non parametric maximum likelihood estimator
2. It may account for data sets with ties, i.e. d_j units failing at the same time t_i, $i = 1, 2, ..., n$:

$$\hat{R}(t) = \prod_{j \in \Gamma} \frac{n_j - d_j}{n_j}$$

3. It is a consistent estimator of $R(t)$ under quite general conditions, with estimated asymptotic variance

$$var\left[\hat{R}(t)\right] = \left[\hat{R}(t)\right]^2 \sum_{j \in \Gamma} \frac{d_j}{n_j\left(n_j - d_j\right)}$$

4. It has an asymptotic normal distribution, since it is a maximum likelihood estimator. Hence, confidence limits for $R(t)$ can be determined using the normal approximation.

Example 9.6 [11]

Suppose that a test has been carried out as described above, with $n = 16$, and the observed lifetimes are (given in months):

31.7	39.2	57.5	65.0	65.8	70.0	75.0	75.2
87.7	88.3	94.2	101.7	105.8	109.2	110.0	130.0

The empirical distribution function $\hat{F}_n(t)$ is illustrated in Fig. 9.24. The empirical survivor function $\hat{R}(t) = 1 - \hat{F}_n(t)$ is illustrated in Fig. 9.25.

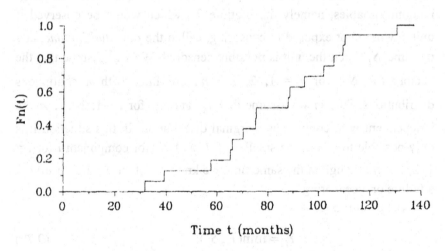

Fig. 9.24: Empirical distribution function $\hat{F}_n(t)$

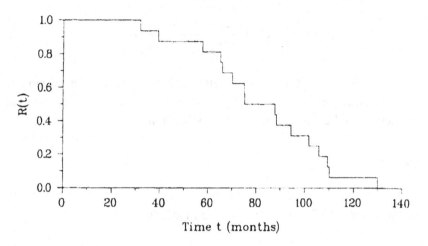

Fig. 9.25: Empirical survival function $\hat{R}(t) = 1 - \hat{F}_n(t)$

Next, let us show how to estimate $R(t)$ from an incomplete data set with censoring of type IV (Section 9.1.3). A set of n numbered units are activated at time $t = 0$, and the censoring time for unit i, S_i is stochastic. Associated with unit i for $i = 1, 2, ..., n$ are two nonnegative random variables, namely the lifetime T_i which would be observed if unit i where not exposed to censoring, called the *potential lifetime*, and the time S_i when the unit is possibly censored. We will assume that the vectors (T_i, S_i) for $i = 1, 2, ..., n$, are i.i.d. with a continuous distributions. Further we assume that T_i and S_i for $i = 1, 2, ..., n$ are independent with continuous marginal distribution. In this situation it is only possible to record the smaller of T_i and S_i for component i for $i = 1, 2, ..., n$, though at the same time we know whether we are observing a failure or a censoring.

Let us introduce

$$Y_i = \min(T_i, S_i) \tag{9.70}$$

and the indicators

$$\delta_i = \begin{cases} 1 & \text{if } T_i \le S_i \\ 0 & \text{if } T_i > S_i \end{cases} \qquad i = 1, 2, ..., n \qquad (9.71)$$

After the life test is terminated, we are left with the data set

$$(Y_1, \delta_1), (Y_2, \delta_2), ..., (Y_n, \delta_n) \qquad (9.72)$$

The following Kaplan-Meier estimation procedure can be applied. Fix $t > 0$ and let $t_{(1)} < t_{(2)} < ... < t_{(n)}$ denote the recorded functioning times, either until failure or to censoring, ordered according to size. Let J_t denote the set of all indices j where $t_{(j)} \le t$ and $t_{(j)}$ represents a failure time. Let n_j denote the number of units, functioning and in observation immediately before time $t_{(j)}$, $j = 1, 2, ..., n$. Then, the Kaplan-Meier estimator of $R(t)$ is defined as

$$\hat{R}(t) = \prod_{j \in J_t} \frac{n_j - 1}{n_j} \qquad (9.73)$$

Example 9.7 [11]

We change the situation given in Example 9.6 so that only the recorded lifetimes which are not starred (*) in Table 9.3 represent times to failure. In Table 9.4, the Kaplan-Meier estimate is determined as a function of time. In the time interval (0, 31.7) until the first failure, it is reasonable to set $\hat{R}(t) = 1$. The estimate is displayed graphically by a Kaplan-Meier plot in Fig. 9.26.

Table 9.3: Computation of the Kaplan-Meier Estimate [11]

Rank j	Inverse Rank $n-j+1$	Ordered Failure and Censoring Times t_j	$\hat{p}_j$	$\hat{R}(t_{(j)})$
0	-	-	1	1.000
1	16	31.7	15/16	0.938
2	15	39.2	14/15	0.875
3	14	57.2	13/14	0.813
4	13	65.0*	1	0.813
5	12	65.8	11/12	0.745
6	11	70.0	10/11	0.677
7	10	75.0*	1	0.677
8	9	75.2*	1	0.677
9	8	87.5*	1	0.677
10	7	88.3*	1	0.677
11	6	94.2*	1	0.677
12	5	101.7*	1	0.677
13	4	105.8	¾	0.508
14	3	109.2*	1	0.508
15	2	110.0	½	0.254
16	1	130.0*	1	0.254

Note: Censoring times are starred(*)

Table 9.4: Kaplan-Meier Estimate as a Function of Time [11]

t	$\hat{R}(t)$
$0 \le t < 31.7$	=1
$31.7 \le t < 39.2$	15/16=0.938
$39.2 \le t < 57.5$	15/16·14/15=0.875
$57.5 \le t < 65.8$	15/16·14/15·13/14=0.813
$65.8 \le t < 70.0$	15/16·14/15·13/14·11/12=0.745
$70.0 \le t < 105.8$	15/16·14/15·13/14·11/12·10/11=0.677
$105.8 \le t < 110.0$	15/16·14/15·13/14·11/12·10/11·3/4=0.508
$110.0 \le t$	15/16·14/15·13/14·11/12·10/11·3/4·1/2=0.254

We see from the equation (9.73) for the Kaplan-Meier estimator and also from Figure 9.26 that $\hat{R}(t)$ is a step function, continuous from the right, that equals 1 at t=0. $\hat{R}(t)$ drops by a factor of $(n_j - 1)/n_j$ at each failure time t_j. The estimator $\hat{R}(t)$ does not change at the censoring times. The effect of the censoring is, however, influencing the values of n_j and hence the size of the steps in $\hat{R}(t)$.

A slightly problematic point is that $\hat{R}(t)$ never reduces to zero when the last time $t_{(n)}$ recorded is a censored time. For this reason $\hat{R}(t)$ is usually taken to be undefined for $t > t_{(n)}$.

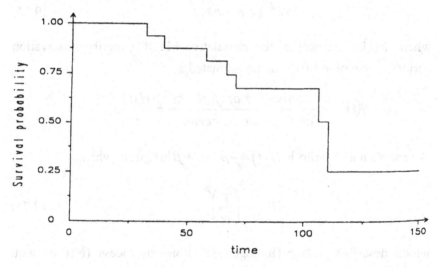

Fig. 9.26: Kaplan-Meier plot of the estimated survival probability $\hat{R}(t)$ [11]

9.5   Reliability growth

It is common practice, during the development of a system, to make engineering changes as the program develops. These changes are generally made in order to correct design deficiencies and thereby to increase reliability. This elimination of design weakness is known as the reliability growth.

Reliability growth can be characterized by [12]:

1. Expressing the cumulative number of failures as a function of operating time.
2. Expressing failure rate as a function of operating time
3. Expressing mean time between failures as a function of time.

A commonly used reliability growth model is the Duane model [10]. Using data from the development programs of several different and complex equipments, Duane observed that the logarithms of observed cumulative MTBFs, $\dfrac{1}{\vartheta(t)}$, was a linear function of time:

$$\ln \vartheta(t) = a + b \ln t \qquad (9.74)$$

where $\vartheta(t)$ reciprocal of the cumulative MTBF over the observation period of operation $[0, t]$, can be estimated as

$$\vartheta(t) = \frac{Total\ number\ of\ failures}{Total\ operating\ period} = \frac{H(t)}{t}.$$

We can then also write $\ln H(t) = -\beta \ln \alpha + \beta \ln t$, from which,

$$H(t) = \left(\frac{t}{\alpha}\right)^{\beta} \qquad (9.75)$$

which describes a Non-Homogeneous Poisson-Process (NHPP) with Weibull intensity

$$h(t) = H'(t) = \frac{\beta t^{\beta-1}}{\alpha^{\beta}} \qquad (9.76)$$

The function $h(t)$ has the same functional form of the instantaneous hazard rate of the Weibull distribution. However, while the instantaneous hazard rate is the conditional probability of failure at $t + \Delta t$ given that there was no failure prior to t (Section 4.5.3), the present intensity function $h(t)$ represents the unconditional probability of failure at time $t + \Delta t$.

We then have:

$$\frac{1}{h(t)} = \text{current MTBF at time } t = \frac{\alpha^\beta}{\beta t^{\beta-1}} \qquad (9.77)$$

$$\phi(t) = \frac{t}{H(t)} = \text{cumulative MTBF for the observation period } [0,t] = \frac{\alpha^\beta}{t^{\beta-1}}$$
$$(9.78)$$

$$\ln \phi(t) = \beta \ln \alpha - (\beta - 1) \ln t \qquad (9.79)$$

When the cumulative MTBF is plotted against the operating time on a log-log paper, it falls on a straight line $\ln \phi(t) = m \ln t + q$ with:

$$m = \text{Slope} = 1 - \beta, \text{ growth factor}$$
$$q = \text{Intercept} = \beta \ln \alpha$$

The two cases which may occur are:

$$\beta < 1 \quad \Rightarrow \quad \textit{reliability growth}$$
$$\beta > 1 \quad \Rightarrow \quad \textit{reliability degradation}$$

The parameters of the Duane growth model, α and β, can be determined either by estimation methods such as the maximum likelihood or by graphical methods. Implicit in the model is the assumption that after breakdown the system is returned to a state identical to that immediately prior to failure.

9.5.1 Maximum likelihood estimation

Direct application of the maximum likelihood estimation on the failure observation data leads to the following estimates of the parameters α and β of the Duane Model:

$$\hat{\beta} = \frac{n}{\sum_{i=1}^{n} ln\left(\frac{T}{t_i}\right)}$$ (9.80)

$$\hat{\alpha} = \frac{T}{n^{(1/\beta)}}$$ (9.81)

n = total number of failures

t_i = failure time, $i = 1, 2, ..., n$

$T = t_o$ *if the test terminates at time* $t_o > t_n$; $\quad$ t_n *if the test terminates at time* t_n

9.5.2   Least Square estimation

For the generic linear relation $y = mx + q$, the Least-Square estimation method applied to the n observation pairs (x_i, y_i), $i=1,2,...,n$, leads to the following estimates:

$$m = \frac{n \cdot \sum_{i=1}^{n} x_i y_i - \left[\sum_{i=1}^{n} x_i\right]\left[\sum_{i=1}^{n} y_i\right]}{n \cdot \sum_{i=1}^{n} x_i^2 - \left[\sum_{i=1}^{n} x_i\right]^2}$$

(9.82)

$$q = \frac{\left[\sum_{i=1}^{n} x_i^2\right]\left[\sum_{i=1}^{n} y_i\right] - \left[\sum_{i=1}^{n} x_i y_i\right]\left[\sum_{i=1}^{n} x_i\right]}{n \cdot \sum_{i=1}^{n} x_i^2 - \left[\sum_{i=1}^{n} x_i\right]^2}$$

Example 9.8 [12]

Consider the data given in Table 9.5.

Table 9.5: Reliability performance data.

(1)	(2)	(3)	(4)	(5)	(6)
Month of Operation	Hours of Operation	Cumulative Hours t	No. of failures	Cumulative Number of failures $H(t)$	Cumulative MTBF $t/H'(t)$
1	541	541	3	3	180.3
2	1171	1712	5	8	214.0
3	1939	3651	4	12	304.3
4	2403	6054	1	13	465.7
5	1718	7772	2	15	518.1
6	2206	9978	2	17	586.9
7	1366	11244	3	20	562.2
8	1529	12873	0	20	643.7
9	1449	14322	2	22	651.0
10	1451	15773	2	24	657.2

To carry out a Duane analysis, the failure data can be arranged as in columns 3 and 5 of Table 9.5. The cumulative MTBF in column 6 is obtained by dividing the total number of failures by the total hours of operation. Using the data from columns 3 and 6, Table 9.6 can be constructed for the Least Square estimation of the slope and intercept parameters of the Duane model (Eqs. 9.82).

Table 9.6: Reliability data for Least-Square estimation

Month No.	*In* cum. Hours of Operation $\ln t$	*In* cum. MTBF $\ln\left(\dfrac{t}{H(t)}\right)$		
	x	y	$x \cdot y$	x^2
1	6.29	5.20	32.71	39.56
2	7.45	5.37	40.00	55.50
3	8.20	5.72	46.90	67.24
4	8.71	6.14	53.48	75.86
5	8.96	6.25	56.00	80.28
6	9.21	6.38	58.76	84.82
7	9.33	6.33	59.06	87.05
8	9.46	6.47	61 .21	89.49
9	9.57	6.48	62.01	91.59
10	9.67	6.49	62.76	93.51
Σ	86.85	60.83	532.89	764.90

Then,

$$n\sum_{i=1}^{n} x_i^2 - \left[\sum_{i=1}^{n} x_i\right]^2 = 10 \cdot 764.90 - 86.85^2 = 7649 - 7543 = 106$$

$$m = \frac{10 \cdot 532.98 - 86.85 \cdot 60.83}{106} = \frac{45.8}{106} = 0.432$$

$$q = \frac{764.90 \cdot 60.83 - 532.89 \cdot 86.85}{106} = \frac{247.37}{106} = 2.33$$

The Least-Square estimates of α and β can be calculated as:

$$m = 1 - \hat{\beta} = 0.432 \implies \hat{\beta} = 1 - 0.432 = 0.568$$

$$q = \hat{\beta}\ln\alpha \implies \hat{\alpha} = e^{\frac{2.33}{0.568}} = 60.47$$

The growth factor $1 - \hat{\beta}$ is then 0.432. The current MTBF at time t is given by:

$$\frac{1}{h(t)} = \frac{\alpha^{\beta}}{\beta t^{\beta-1}}$$

and the current failure rate at time t is given by

$$h(t) = \frac{\beta t^{\beta-1}}{\alpha^{\beta}}.$$

For example, the current failure rate at the end of 1 month of operation (541 hours) is:

$$h(541) = \frac{0.568 \cdot 541^{-0.432}}{60.47^{0.568}} = 0.05526 \cdot 541^{-0.432} = 3.645 \frac{failures}{10^3 \, hours}$$

and the current failure rate at the end of 10 months of operation (i.e., 15773 hours) is:

$$h(15773) = 0.05526 \cdot 15773^{-0.432} = 0.849 \frac{failures}{10^3 \, hours}$$

Thus, there is an improvement of $\dfrac{3.645 - 0.849}{3.645} \cdot 100 = 76.7\%$.

The Duane plot with estimated current MTBF and cumulative MTBF lines are given in Fig. 9.27 [9].

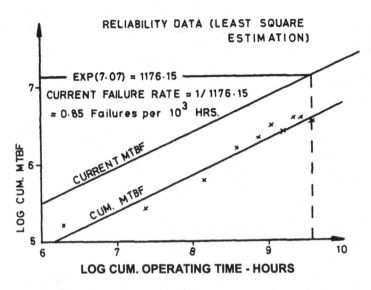

Fig. 9.27: Duane plot for the least square estimation [9]

Example 9.9 [12]

The observed number of failures in a certain steam injection system on an oil production platform are given in Table 9.7. Failure of this system causes production stoppage, whose cost is estimated at £10,000 on average (this includes cost of lost production and labour). The cost of complete overhaul including lost production, replacement and labour is estimated to be £50,000. What is the optimum overhaul policy?

Table 9.7: Observed failure for steam injection system

Month of Operation	Observed Number of Failures
1	0
2	1
3	2
4	5
5	7
6	12
7	16

Solution:

Table 9.8 and Table 9.9 report the data for the Duane model parameters estimation by Least-Square. The number n is equal to 6 since there are only 6 intervals with failures.

Table 9.8: Optimum overhaul policy for steam injection system

Month of Operation	Observed Number of Failures	Cumulative Number of Failures	Cumulative MTBF (months)
1	0	0	-
2	1	1	0.05
3	2	3	1.00
4	5	8	0.50
5	7	15	0.33
6	12	27	0.22
7	16	43	0.163

Table 9.9: Least Square estimation data

No.	x	y	xy	x^2
1	-	-	-	-
2	0.69	0	0	0.48
3	1.10	0	0	0.21
4	1.39	-0.69	-0.96	1.93
5	1.60	-1.11	-1.78	2.56
6	1.79	-1.51	-2.70	3.20
7	1.95	-1.81	-3.53	3.80
$\sum$	8.52	-5.12	-8.97	13.18

The parameters estimates are (Eqs. 9.82):

$$Slope = m = \frac{6 \cdot (-8.97) - 8.52(-5.12)}{6*13.18 - 8.52^2} = \frac{-53.82 + 43.62}{10.2} = \frac{-10.2}{10.2} = -1$$

$$Intercept = q = \frac{13.8 \cdot (-5.12) - (-8.97) \cdot 8.52}{10.2} = \frac{-70.66 + 76.42}{10.2} = \frac{5.76}{10.2} = 0.565$$

from which,

$$\hat{\beta} = 1 - m = 1 - (-1) = 2$$

$$\hat{\alpha} = \exp\left[\frac{q}{\beta}\right] = 1.33.$$

The cumulative number of failures is given by $H(t) = \left(\frac{t}{\alpha}\right)^{\beta} = \left(\frac{t}{1.33}\right)^2$.

Let C_1 be the cost associated with the breakdown and C_2 be the cost associated with overhaul. Then, the total cost of operation for a time period t is given by:

$$C(t) = C_1 H(t) + C_2$$

and the cost per unit operating time is given by:

$$\gamma(t) = \frac{C(t)}{t} = \frac{C_1 H(t) + C_2}{t}$$

where $\gamma(t)$ is the cost/unit operating time,

$$\gamma(t) = \frac{C_1\left(\frac{t}{\alpha}\right)^{\beta} + C_2}{t}$$

In order to minimize $\gamma(t)$, one has to solve the following equation

$$\frac{d\gamma}{dt} = 0$$

$$\frac{C_1(\beta-1)\, t^{\beta-2}}{\alpha^\beta} - \frac{C_2}{t^2} = 0 \quad \Rightarrow \quad \frac{C_1(\beta-1)\, t^{\beta-2}}{\alpha^\beta} = \frac{C_2}{t^2}$$

or $\qquad t^\beta = \dfrac{\alpha^\beta C_2}{C_1(\beta-1)} = \dfrac{\alpha^\beta \,{C_2}/{C_1}}{(\beta-1)} \quad \Rightarrow$

$$t^* = \alpha\left[\frac{\alpha^\beta C_2}{C_1(\beta-1)}\right]^{1/\beta}$$

For the given data:

$C_1 = 10000; \qquad C_2 = 50000; \qquad C_2/C_1 = 5; \qquad \beta = 2; \quad \alpha = 1.33;$

and the optimal overhaul time is:

$$t^* = 1.33\left[\frac{5}{1}\right]^{1/2} = 2.97 \text{ months}$$

The equipment should be overhauled approximately once in every three months. The corresponding cost/unit operating time is given by

$$\gamma(3) = \frac{10000\cdot\left(3/1.33\right)^2 + 50000}{3} = \frac{£33626}{month}$$

Since it is not always practicable to schedule the overhaul at the exact optimum, it is useful to examine the sensitivity of the cost curve with respect to the time between overhauls (Table 9.10 and Fig. 9.28).

Table 9.10: Cost Curve

Month of Operation	Expected No. of Failures ($H(t)$)	Running cost $C_1 H(t) + C_2$ (£000)	Cost per Unit Time (£000)
0	-	-	-
1	0.57	55.7	55.7
2	2.26	72.6	36.3
3	5.09	100.9	33.6
4	9.05	140.5	35.1
5	14.10	191.0	38.2
6	20.35	253.5	42.3
7	27.70	327.0	46.7

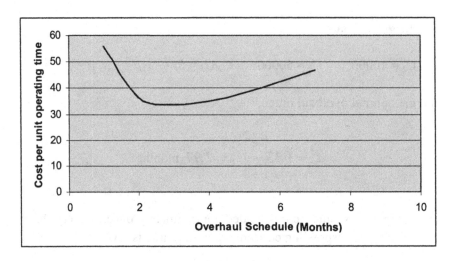

Fig. 9.28: Optimum overhaul policy for the steam injection system

9.6   Reliability prediction from stress-strength models

9.6.1   Introduction

The prediction of reliability from failure statistics does not concern itself with what happens inside the unit, although it is intuitive that a unit fails when the stress imposed exceeds the strength. To properly describe such failure mechanism quantitatively, stress-strength models need to be introduced [13].

The stress on a unit is the total sum of internal (created by the operational use) and external stresses (imposed by the environmental conditions of use).

Obviously, the stress and strength operating on identical units are not fixed quantities and vary from unit to unit even if the best quality control procedures are used. Therefore, stress and strength should be considered as random variables. In the following, an outline is given of the basis for computing the component reliability from the knowledge of the distributions associated with these random variables.

In the past, the concept of safety factors has been widely used in the design of engineering systems:

$$safety\ factor = \frac{ultimate\ strength}{working\ stress}$$

where the ultimate strength and the working stress are considered as fixed known values, with no consideration given to their variability.

To illustrate this concept, let us assume that the stress or load L applied to a component is normally distributed with density function $f_L(l)$ characterized by a mean μ_L and a standard deviation σ_L (Fig. 9.29). The strength of the unit has been determined to be S_1: thus, the unit may fail only when the stress acting upon it is greater than S_1. The probability of failure is then given by [13]:

$$F = P[L > S_1] = \int_{S_1}^{\infty} f_L(l)dl \qquad (9.83)$$

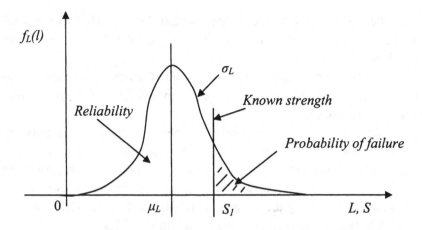

Fig. 9.29: Reliability from Stress-Strength Model

Equation (9.83) gives the unreliability of a unit whose strength is a known, invariant value S_l. When the variability of both stress and strength is taken into considerations, the probability of failure of a unit depends on the area of overlap between the distributions of the stress and strength random variables (Fig. 9.30).

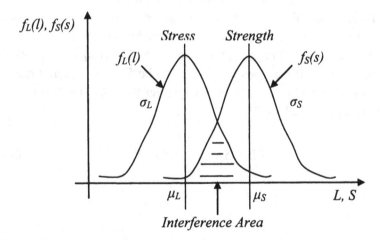

Fig. 9.30: Probability of Failure from Stress-Strength Distributions

To enhance the reliability of a unit, one can then (Figs. 9.31)

a)   Shift apart the stress and strength distributions
b)   Reduce the variability of stress by a better regulation of stress and control of the environmental conditions
c)   Reduce the variability of strength by stricter quality control during the production phase.

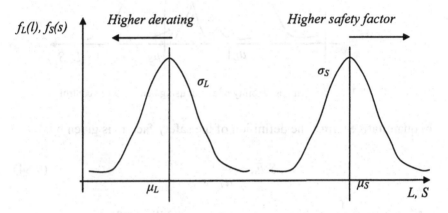

Fig. 9.31a: Improving Reliability of a Unit using derating and safety factors

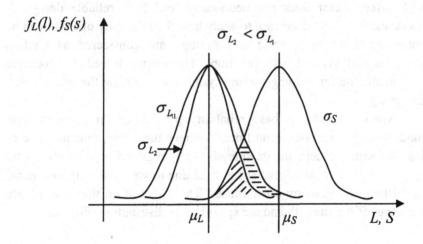

Fig. 9.31b: Improving Reliability of a Unit using stress regulation

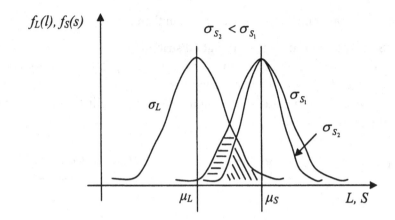

Fig. 9.31c: Improving Reliability of a Unit using better quality control

In quantitative terms, the definition of the safety factors is given by:

$$S.F. = \frac{S_1}{\mu_L} \tag{9.84}$$

where no account is given to the variability of stress since no consideration is given to μ_L. However, contrary to the general belief, a high safety factor does not necessarily lead to a reliable design. It becomes, therefore, necessary to study how the reliability of a unit can be computed when both stress and strength are considered as random variables with given density functions. This approach leads to economic and reliable designs of engineering systems, eliminating the risk of over-designing.

Another reason to base reliability predictions on stress-strength models is the dependence on time. Whereas the stress remains more or less the same, except for its spread, the strength of a unit varies with time. Usually strength decreases in time due to wear and tear, corrosion, metal fatigue and many other causes. The effect of all this is to reduce the mean of the strength and the spread of its distribution (Fig. 9.32).

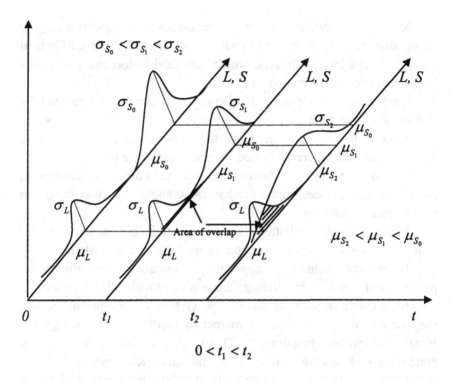

$$0 < t_1 < t_2$$

Fig. 9.32: Degradation of Strength of a Unit Over Time

9.6.2 Internal and external causes of stress

The increase of the stress in a unit can be due to either internal or external causes. One of the major unreliability causes is the poor thermal transfer. In the majority of cases, the fault does not lie in the design but in not having given adequate attention to the heat generated within the components. If adequate cooling of the components inside the equipment has not been planned, the component might experience a building up of hot spots leading to higher localized stresses. This may even be due to maintenance personnel's failing to replace clocked air filters. In any case, whatever the cause, poor heat transfer may lead to physical damage or accelerated chemical reactions, which affect the material's properties.

Similarly, a sudden change in temperature or rapidly changing temperature cycles produce additional stresses on components. Effects of thermal shocks induce stresses, which may lead to loosening of solder joints, cracking, delamination, etc.

Furthermore, the natural frequency of components must not be overlooked since resonance may occur if the natural frequency is within the vibration frequency range. Once the resonance occurs it can cause deflection and may increase stresses beyond acceptable limits.

Excessive vibrations themselves, due to worn out bearings or misalignments, can deteriorate mechanical strength and cause fatigue or overstress in components.

Electromagnetic radiations and electrostatic discharges can also cause excessive electrical stresses on components and subsystems.

In nuclear technology applications, radiation also affects the properties of materials by altering their atomic or molecular structure.

Besides these internal factors, a system or equipment is also subjected to many external environmental factors, which may greatly impair its proper functioning. These factors depend on climatic conditions such as daily maximum and minimum temperatures and their variations, on altitude, on rain, on humidity, on sand or dust, on atmospheric salinity, etc. All these factors increase the stress in the components. Temperature and humidity are major factors.

9.6.3   Physics of failures

As mentioned before, the strength S of a component is a random variable and varies not only from batch to batch of the production line of a given unit but also decreases with time. Thus, the failure occurs when overtime the strength becomes less than the applied load. To derive the reliability model we must then know the law of decrease in time of the strength, $S(t)$, and its initial value, S_0.

9.6.4 Reliability from stress-strength distributions

The reliability of a unit is the probability that the strength S is greater than the stress or load L for all possible values of the load L [13]:

$$R = \int_{-\infty}^{\infty} f_L(l) \underbrace{\left[\int_{l}^{\infty} f_s(s)\,ds \right]}\,dl \tag{9.85}$$

Probability of the
strength $S > l = 1 - F_S(l)$

or

$$R = \int_{-\infty}^{\infty} f_S(s) \underbrace{\left[\int_{-\infty}^{s} f_L(l)\,dl \right]}\,ds \tag{9.86}$$

Probability of the load
$L < s = F_L(s)$

If we express $\Delta = S - L$, then $S = \Delta + L$ and

$$R = P[\Delta \geq 0] = \int_{0}^{\infty} f_\Delta(\delta)\,d\delta$$

$$= \int_{0}^{\infty} \underbrace{\left[\int_{0}^{\infty} f_s(\delta + l) f_L(l)\,dl \right]}\,d\delta \tag{9.87}$$

$$f_s(\delta)$$

References

[1]   Henley, E.J. and Kumamoto, H., Probabilistic risk assessment, NY, IEEE Press, 1992.

[2]   Mann, N.R., R. E. Schafer, and N. D. Singpurwalla, Methods for Statistical Analysis of Reliability and Life Data., New York, Wiley, 1974.

[3]   Kalbfleisch, J. D., and R. L. Prentice., The Statistical Analysis of Failure Time Data, New York, Wiley, 2002.

[4]   Lawless, J. F., Statistical Models and Methods for Lifetime Data, New York, Wiley, 1982.

[5]   Jensen, F., and N. E. Petersen, Burn–in: An Engineering Approach to the Design and Analysis of Burn-in Procedures., New York, Wiley, 1982.

[6]   Cox, D. R., and D. Oakes, Analysis of Survival Data., London, Chapman and Hall, 2003.

[7]   Viertl, R., Statistical Methods in Accelerated Life Testing., Vandenhoeck & Ruprecht. Göttingen, 1988.

[8]   Nelson W., Accelerated Testing: Statistical Models, Test Plans, and Data Analyses, New York, Wiley, 1990.

[9]   Ang, A.H. and Tang, W.H., Probability Concepts in Engineering Planning and Design. Vol. 1: Basic Principles, 1975.

[10] Duane, J.T., Learning Curve Approach To Reliability Monitoring", IEEE Transactions on Aerospace, Vol.2, 1964.

[11] Rausand, M. and Hoyland, A., System Reliability Theory, Wiley, 2004.

[12] Keller, A.Z., Reliability Aging and Growth Modeling, In Reliability Modeling and Applications, A.G. Colombo and A.Z. Keller Eds., Kluwer, 1986.

[13] Misra, K.B., Reliability Analysis and Prediction, Elsevier, 1992.

Appendix A: Table of Standard Normal Cumulative Distribution

$$F(\xi) = \frac{1}{\sqrt{2\pi}} \int_{-\infty}^{\xi} e^{-\frac{1}{2}x^2} dx$$

ξ	$F(\xi)$	ξ	$F(\xi)$	ξ	$F(\xi)$
0.00	0.500000	0.50	0.691463	1.00	0.841345
0.01	0.503989	0.51	0.694975	1.01	0.843752
0.02	0.507978	0.52	0.698468	1.02	0.846136
0.03	0.511966	0.53	0.701944	1.03	0.848495
0.04	0.515954	0.54	0.705401	1.04	0.850830
0.05	0.519939	0.55	0.708840	1.05	0.853141
0.06	0.523922	0.56	0.712260	1.06	0.855428
0.07	0.527904	0.57	0.715661	1.07	0.857690
0.08	0.531882	0.58	0.719043	1.08	0.859929
0.09	0.535857	0.59	0.722405	1.09	0.862143
0.10	0.539828	0.60	0.725747	1.10	0.864334
0.11	0.543796	0.61	0.729069	1.11	0.866500
0.12	0.547759	0.62	0.732371	1.12	0.868643
0.13	0.551717	0.63	0.735653	1.13	0.870762
0.14	0.555671	0.64	0.738914	1.14	0.872857
0.15	0.559618	0.65	0.742154	1.15	0.874928
0.16	0.563500	0.66	0.745374	1.16	0.876976
0.17	0.567494	0.67	0.748572	1.17	0.878999
0.18	0.571423	0.68	0.751748	1.18	0.881000
0.19	0.575345	0.69	0.754903	1.19	0.882977
0.20	0.579260	0.70	0.758036	1.20	0.884930
0.21	0.583166	0.71	0.761148	1.21	0.886860
0.22	0.587064	0.72	0.764238	1.22	0.888767
0.23	0.590954	0.73	0.767305	1.23	0.890651
0.24	0.549835	0.74	0.770350	1.24	0.892512
0.25	0.598706	0.75	0.773373	1.25	0.894350
0.26	0.602568	0.76	0.776373	1.26	0.896165
0.27	0.606420	0.77	0.779350	1.27	0.897958
0.28	0.610262	0.78	0.782305	1.28	0.899727
0.29	0.614092	0.79	0.785236	1.29	0.901475

ξ	$F(\xi)$	ξ	$F(\xi)$	ξ	$F(\xi)$
0.30	0.617912	0.80	0.788145	1.30	0.903199
0.31	0.621720	0.81	0.791030	1.31	0.904902
0.32	0.623517	0.82	0.793892	1.32	0.906583
0.33	0.629301	0.83	0.796731	1.33	0.908241
0.34	0.633072	0.84	0.799546	1.34	0.909877
0.35	0.636831	0.85	0.802337	1.35	0.911492
0.36	0.640576	0.86	0.805105	1.36	0.913085
0.37	0.644309	0.87	0.807850	1.37	0.914656
0.38	0.648027	0.88	0.810570	1.38	0.916207
0.39	0.651732	0.89	0.813267	1.39	0.917735
0.40	0.655422	0.90	0.815940	1.40	0.919243
0.41	0.659097	0.91	0.818589	1.41	0.920730
0.42	0.662757	0.92	0.821214	1.42	0.922196
0.43	0.666402	0.93	0.823815	1.43	0.923641
0.44	0.670032	0.94	0.826391	1.44	0.925066
0.45	0.673645	0.95	0.828944	1.45	0.926471
0.46	0.677242	0.96	0.831473	1.46	0.927855
0.47	0.680823	0.97	0.833977	1.47	0.929219
0.48	0.684387	0.98	0.836457	1.48	0.930563
0.49	0.687933	0.99	0.838913	1.49	0.931888
1.50	0.933193	2.00	0.977250	2.50	0.993790
1.51	0.934478	2.01	0.977784	2.51	0.993963
1.52	0.935744	2.02	0.978308	2.52	0.994132
1.53	0.936992	2.03	0.978822	2.53	0.994267
1.54	0.938220	2.04	0.979325	2.54	0.994457
1.55	0.939429	2.05	0.979818	2.55	0.994614
1.56	0.940620	2.06	0.980301	2.56	0.994766
1.57	0.941792	2.07	0.980774	2.57	0.994915
1.58	0.942947	2.08	0.981237	2.58	0.995060
1.59	0.944083	2.09	0.981691	2.59	0.995201
1.60	0.945201	2.10	0.982136	2.60	0.995339
1.61	0.946301	2.11	0.982571	2.61	0.995473
1.62	0.947384	2.12	0.982997	2.62	0.995604
1.63	0.948449	2.13	0.983414	2.63	0.995731
1.64	0.949497	2.14	0.983823	2.64	0.995855

ξ	$F(\xi)$	ξ	$F(\xi)$	ξ	$F(\xi)$
1.65	0.950529	2.15	0.984223	2.65	0.995975
1.66	0.951543	2.16	0.984614	2.66	0.996093
1.67	0.952540	2.17	0.984997	2.67	0.996207
1.68	0.953521	2.18	0.985371	2.68	0.996319
1.69	0.954486	2.19	0.985738	2.69	0.996427
1.70	0.955435	2.20	0.986097	2.70	0.996533
1.71	0.956367	2.21	0.986447	2.71	0.996636
1.72	0.957284	2.22	0.986791	2.72	0.996736
1.73	0.958185	2.23	0.987126	2.73	0.996833
1.74	0.959071	2.24	0.987455	2.74	0.996928
1.75	0.959941	2.25	0.987776	2.75	0.997020
1.76	0.960796	2.26	0.988089	2.76	0.997110
1.77	0.961636	2.27	0.988396	2.77	0.997197
1.78	0.962426	2.28	0.988696	2.78	0.997282
1.79	0.963273	2.29	0.988989	2.79	0.997365
1.80	0.964070	2.30	0.989276	2.80	0.997445
1.81	0.964852	2.31	0.989556	2.81	0.997523
1.82	0.965621	2.32	0.989830	2.82	0.997599
1.83	0.966375	2.33	0.990097	2.83	0.997673
1.84	0.967116	2.34	0.990358	2.84	0.997744
1.85	0.967843	2.35	0.990613	2.85	0.997814
1.86	0.968557	2.36	0.990863	2.86	0.997882
1.87	0.969258	2.37	0.991106	2.87	0.997948
1.88	0.969946	2.38	0.991344	2.88	0.998012
1.89	0.970621	2.39	0.991576	2.89	0.998074
1.90	0.971284	2.40	0.991802	2.90	0.998134
1.91	0.971933	2.41	0.992024	2.91	0.998193
1.92	0.972571	2.42	0.992240	2.92	0.998250
1.93	0.973197	2.43	0.992451	2.93	0.998305
1.94	0.973810	2.44	0.992656	2.94	0.998359
1.95	0.974412	2.45	0.992857	2.95	0.998411
1.96	0.975002	2.46	0.993053	2.96	0.998462
1.97	0.975581	2.47	0.993244	2.97	0.998511
1.98	0.976148	2.48	0.993431	2.98	0.998559
1.99	0.976705	2.49	0.993613	2.99	0.998605

ξ	$F(\xi)$	ξ	$F(\xi)$	ξ	$1-F(\xi)$
3.00	0.998630	3.50	0.999767	4.00	0.316712E-04
3.01	0.998694	3.51	0.999776	4.05	0.256088E-04
3.02	0.998736	3.52	0.999784	4.10	0.206575E-04
3.03	0.998777	3.53	0.999792	4.15	0.166238E-04
3.04	0.998817	3.54	0.999800	4.20	0.133458E-04
3.05	0.998856	3.55	0.999807	4.25	0.106883E-04
3.06	0.998893	3.56	0.999815	4.30	0.853906E-05
3.07	0.998930	3.57	0.999821	4.35	0.680688E-05
3.08	0.998965	3.58	0.999828	4.40	0.541234E-05
3.09	0.998999	3.59	0.999835	4.45	0.429351E-05
3.10	0.999032	3.60	0.999841	4.50	0.339767E-05
3.11	0.999065	3.61	0.999847	4.55	0.268230E-05
3.12	0.999096	3.62	0.999853	4.60	0.211245E-05
3.13	0.999126	3.63	0.999858	4.65	0.165968E-05
3.14	0.999155	3.64	0.999864	4.70	0.130081E-05
3.15	0.992184	3.65	0.999869	4.75	0.101708E-05
3.16	0.999119	3.66	0.999874	4.80	0.793328E-06
3.17	0.999238	3.67	0.999879	4.85	0.617307E-06
3.18	0.999264	3.68	0.999883	4.90	0.479183E-06
3.19	0.999289	3.69	0.999888	4.95	0.371067E-06
3.20	0.999313	3.70	0.999892	5.00	0.286652E-06
3.21	0.999336	3.71	0.999806	5.10	0.169827E-06
3.22	0.999359	3.72	0.999900	5.20	0.996443E-07
3.23	0.999381	3.73	0.999904	5.30	0.579013E-07
3.24	0.999402	3.74	0.999908	5.40	0.333204E-07
3.25	0.999423	3.75	0.999912	5.50	0.189896E-07
3.26	0.999443	3.76	0.999915	5.60	0.107176E-07
3.27	0.999462	3.77	0.999918	5.70	0.599037E-08
3.28	0.999481	3.78	0.999922	5.80	0.331575E-08
3.29	0.999499	3.79	0.999925	5.90	0.181751E-08
3.30	0.999516	3.80	0.999928	6.00	0.986588E-09
3.31	0.999533	3.81	0.999931	6.10	0.530343E-09
3.32	0.999550	3.82	0.999933	6.20	0.282316E-09
3.33	0.999566	3.83	0.999936	6.30	0.148823E-09
3.34	0.999581	3.84	0.999938	6.40	0.77688 E-10

ξ	$F(\xi)$	ξ	$F(\xi)$	ξ	$1-F(\xi)$
3.35	0.999596	3.85	0.999941	6.50	0.40160 E-10
3.36	0.999610	3.86	0.999943	6.60	0.20558 E-10
3.37	0.999624	3.87	0.999946	6.70	0.10421 E-10
3.38	0.999637	3.88	0.999948	6.80	0.5231 E-11
3.39	0.999650	3.89	0.999950	6.90	0.260 E-11
3.40	0.999663	3.90	0.999952	7.00	0.128 E-11
3.41	0.999675	3.91	0.999954	7.10	0.624 E-12
3.42	0.999687	3.92	0.999956	7.20	0.301 E-12
3.43	0.999698	3.93	0.999958	7.30	0.144 E-12
3.44	0.999709	3.94	0.999959	7.40	0.68 E-13
3.45	0.999720	3.95	0.999961	7.50	0.32 E-13
3.46	0.999730	3.96	0.999963	7.60	0.15 E-13
3.47	0.999740	3.97	0.999964	7.70	0.70 E-14
3.48	0.999749	3.98	0.999966	7.80	0.30 E-14
3.49	0.999758	3.99	0.999967	7.90	0.15 E-14

Appendix B: Table of Chi-Square Cumulative Distribution $\chi_\alpha^2(f)$

f\ α	0.995	0.99	0.975	0.95	0.90	0.10	0.05	0.025	0.01	0.005
1	—	—	0.001	0.004	0.016	2.706	3.841	5.024	6.635	7.879
2	0.010	0.020	0.051	0.103	0.211	4.605	5.991	7.378	9.210	10.597
3	0.072	0.115	0.216	0.352	0.584	6.251	7.815	9.348	11.345	12.838
4	0.207	0.297	0.484	0.711	1.064	7.779	9.488	11.143	13.277	14.860
5	0.412	0.554	0.831	1.145	1.610	9.236	11.070	12.833	15.086	16.750
6	0.676	0.872	1.237	1.635	2.204	10.645	12.592	14.449	16.812	18.548
7	0.989	1.239	1.690	2.167	2.833	12.017	14.067	16.013	18.475	20.278
8	1.344	1.646	2.180	2.733	3.490	13.362	15.507	17.535	20.090	21.955
9	1.735	2.088	2.700	3.325	4.168	14.684	16.919	19.023	21.666	23.589
10	2.156	2.558	3.247	3.940	4.865	15.987	18.307	20.483	23.209	25.188
11	2.603	3.053	3.816	4.575	5.578	17.275	19.675	21.920	24.725	26.757
12	3.074	3.571	4.404	5.226	6.304	18.549	21.026	23.337	26.217	28.300
13	3.565	4.107	5.009	5.892	7.042	19.812	22.362	24.736	27.688	29.819
14	4.075	4.660	5.629	6.571	7.790	21.064	23.685	26.119	29.141	31.319
15	4.601	5.229	6.262	7.261	8.547	22.307	24.996	27.488	30.578	32.801
16	5.142	5.812	6.908	7.962	9.312	23.542	26.296	28.845	32.000	34.267
17	5.697	6.408	7.564	8.672	10.085	24.769	27.587	30.191	33.409	35.718
18	6.265	7.015	8.231	9.390	10.865	25.989	28.869	31.526	34.805	37.156
19	6.844	7.633	8.907	10.117	11.651	27.204	30.144	32.852	36.191	38.582
20	7.434	8.260	9.591	10.851	12.443	28.412	31.410	34.170	37.566	39.997
21	8.034	8.897	10.283	11.591	13.240	29.615	32.671	35.479	38.932	41.401
22	8.643	9.542	10.982	12.338	14.041	30.813	33.924	36.781	40.289	42.796
23	9.260	10.196	11.689	13.091	14.848	32.007	35.172	38.076	41.638	44.181
24	9.886	10.856	12.401	13.848	15.659	33.196	36.415	39.364	42.980	45.559
25	10.520	11.524	13.120	14.611	16.473	34.382	37.652	40.646	44.314	46.928
26	11.160	12.198	13.844	15.379	17.292	35.563	38.885	41.923	45.642	48.290
27	11.808	12.879	14.573	16.151	18.114	36.741	40.113	43.195	46.963	49.645
28	12.461	13.565	15.308	16.928	18.939	37.916	41.337	44.461	48.278	50.993
29	13.121	14.256	16.047	17.708	19.768	39.087	42.557	45.722	49.588	52.336
30	13.787	14.953	16.791	18.493	20.599	40.256	43.773	46.979	50.892	53.672
40	20.707	22.164	24.433	26.509	29.051	51.805	55.758	59.342	63.691	66.766
50	27.991	29.707	32.357	34.764	37.689	63.167	67.505	71.420	76.154	79.490
60	35.534	37.485	40.482	43.188	46.459	74.397	79.082	83.298	88.379	91.952
70	43.275	45.442	48.758	51.739	55.329	85.527	90.531	95.023	100.425	104.215
80	51.172	53.540	57.153	60.391	64.278	96.578	101.879	106.629	112.329	116.321
90	59.196	61.754	65.647	69.126	73.291	107.565	113.145	118.136	124.116	128.299
100	67.328	70.065	74.222	77.929	82.358	118.498	124.342	129.561	135.807	140.169